Austerity Ecology & The Collapse-Porn Addicts

A Defence of Growth, Progress, Industry and Stuff

Austerity Ecology & The Collapse-Porn Addicts

A Defence of Growth, Progress, Industry and Stuff

Leigh Phillips

Winchester, UK
Washington, USA

First published by Zero Books, 2015
Zero Books is an imprint of John Hunt Publishing Ltd., Laurel House, Station Approach,
Alresford, Hants, SO24 9JH, UK
office1@jhpbooks.net
www.johnhuntpublishing.com
www.zero-books.net

For distributor details and how to order please visit the 'Ordering' section on our website.

Text copyright: Leigh Phillips 2014

ISBN: 978 1 78279 960 3
Library of Congress Control Number: 2015930348

A CIP catalogue record for this book is available from the British Library.

Design: Stuart Davies

Printed and bound by CPI Group (UK) Ltd, Croydon, CR0 4YY, UK

We operate a distinctive and ethical publishing philosophy in all
areas of our business, from our global network of authors to
production and worldwide distribution.

CONTENTS

O God, I could be bounded in a nutshell and count myself a king of infinite space, were it not that I have bad dreams.
– *Hamlet, Act 2, Scene 2*

VISUALIZE INDUSTRIAL COLLAPSE
– *Earth First! Bumper-sticker*

Preface

A Mexico City suburb, June 2012

Framed posters of Subcomandante Marcos, with his trademark balaclava and pipe, and of his fellow Zapatista rebels are hung on the walls of her apartment-cum-office.

It's an uncomfortably nostalgic moment for me. I am briefly distracted from the interview I am supposed to be conducting as I think how I used to have Zapatista posters just like them on my walls. The militant indigenous-socialist group from the southern Mexican state of Chiapas had captured the imagination of the activist left in the 1990s via their resistance to the North American Free Trade Agreement (NAFTA) and helped inspire what came to be called the anti-globalisation movement around the world at the turn of the millennium—a movement I had been heavily involved in as a student before I became a journalist.

I refocus and turn back to the interview. I am speaking to Silvia Ribeiro, the Latin American director of the Action Group on Erosion, Technology and Concentration, more commonly known as the "ETC Group", pronounced *et cetera*. ETC is a boutique environmentalist NGO that specialises in critiques of advanced technology, based in Ottawa but with a couple of offices in the Mexican capital and the Philippines. Their substantial international influence belies their small size however, as ETC publications often serve as the foundation of global campaigns mounted by much larger groups such as Greenpeace or Friends of the Earth. The worldwide movement against genetic modification, for example, was largely born of the work of ETC's co-founder, Pat Mooney. They are fierce opponents of the biotechnology sector in general and currently working hard to build international opposition to cloning, the emerging field of synthetic biology, and a suite of medical interventions they are calling 'human enhancement'. Neuroscience, geo-engineering and nanotechnology are also the subjects of sharply critical ETC Group publications. It is this last topic, nanotech, that has led me to seek them out.

Over the course of the previous year, a radical environmentalist group taking for itself the strange moniker of Individuals Tending towards Savagery (*Individualidades tendiendo a lo salvaje*, or ITS, also translatable as 'Individualities Tending towards the Wild') has launched a series of bombings and attempted bombings of nanotechnology researchers at a number of universities in and near Mexico City. Five such mail bombs have exploded or been found and defused. The group for its part claims to have engaged in 11 such attempts. Three researchers and a security guard have been injured—one scientist severely. The group would in 2013 also retroactively claim responsibility for the murder of biotechnology researcher Ernesto Méndez Salinas, shot in the head by a mysterious assailant in 2011. In their communiqués, the ITS denounce the scientists as the vanguard of an industrial civilisation that is killing the planet. They wish for a return to nature, a return to the wild. A few weeks before my arrival in Mexico, a loosely allied Italian group calling itself the 'Olga Cell of the Informal Anarchist Federation - International Revolutionary Front' has kneecapped—that is, shot in the knee—a nuclear engineer in Genoa working for the industrial conglomerate Finmeccanica. The same group sent a letter bomb to the offices of Swiss pro-nuclear lobbyists in 2011 and attempted to bomb IBM's nanotechnology laboratory in Switzerland in 2010. "Finmeccanica means bio- and nanotechnology. Finmeccanica means death and suffering, new frontiers of Italian capitalism," the Olga Cell communiqué read.

As a result of this spate of bizarre attacks targeting scientists, I have been sent to Mexico by *Nature*, the British science journal, to investigate and write a feature on the violence. I speak to a number of academics who have been affected and meet a pair of molecular biologists whose laboratory and that of a neighbouring researcher have twice been targeted by arsonists that they believe to be part of the same eco-anarchist milieu. The ETC Group for its part is not suspected of any links to ITS and has denounced the bombings. Yet ITS makes repeated references to ETC reports on nanotechnology in its statements, and all the nanotech and biotech researchers I speak to are furious at the NGO. When I ask whether they think ETC is connected in any way, one researcher pauses a long time before answering. She then says carefully, diplomatically: "No, not directly, but they have helped created a climate where these things can happen."

The ETC Group want a worldwide moratorium on nanotechnology research and warn that the "likely future threat is that the merger of living and non-living matter will result in hybrid organisms and products that are not easy to control and behave in unpredictable ways". The researchers dismiss such beliefs as crackpot, but worry that these and similar ideas are nonetheless very mainstream, even if most people would not support the violence of a group like ITS.

Ribeiro for her part is very keen to make clear that her group want nothing to do with ITS. The bombings have been a "sick development", she tells me. "These kinds of attacks—they are benefiting the development of nanotechnology," she continues. "It polarized the discussion. Do you want nanotech or the bomb?"

The conversation meanders. While I don't hold the same views on nanotech as ETC, Ribeiro is a trenchant critic of the free market and social injustice, and so am I. Today I may be wearing my science-journalist hat, but I am a lefty myself. For a range of publications, I have written about corporate regulatory capture, lobbyist corruption, and the cruelty of austerity. As a reporter in the European capital for almost a decade, I worked to expose the problems with so many market solutions to global warming, from emissions trading to carbon-capture-and-storage and the fiction of carbon offsets. I've written about the dangers from Arctic oil and gas exploration, the decimation of healthcare systems in eastern Europe, and the hollowing out of democracy in the wake of the Eurozone crisis. I discover that like me, Silvia is hopeful about the recent *Yo Soy 132* mass student movement in Mexico inspired by Occupy Wall Street and the Arab Spring. There's a lot, an awful lot, that we agree on.

I leave contemplating about how familiar Silvia seems to me, how I recognised in her face and her way of talking so many comrades that I've marched with on hundreds of demonstrations and picket lines, or sat next or listened to at left-wing public meetings or conferences over the years.

The same feeling comes over me a few days later. I'm still in Mexico, but now at an anarchist punk and ska show that is raising funds for a fight against jailed political activists. I'm there in the ultimately forlorn hope that I might be able to speak to somebody who knows somebody who knows an ITS member. I've lost count how many political benefit gigs that I've been to. I may never have faced the same level of government and police repression as the young people at the concert, as activists in Mexico, but I've been arrested on a picket line, punched in the face by Italian carabinieri, clubbed by Austrian riot cops and kettled, pepper-sprayed and tear-gassed more times than I can remember. These are my people.

And so I begin to think to myself, my god, how on earth did it happen that environmentalism, the left—anyone with progressive ideas so akin to my own—ended up at such a violent, anti-science, anti-modernist place? A couple of years after I had come home from Mexico, I was furious to find at a local left-wing bookshop a magazine published by activists at the University of Guelph that featured an article celebrating the ITS bombings. A stall at a book fair near where I live that I attended last summer sold a book that had uncritically compiled all the communiqués of ITS.

Of course you can find such barbarity-cheerleading cretins anywhere, and the Mexican bombers are a fringe group, but if you set to one side their terrorism, you have to acknowledge that anti-technology, anti-science and anti-industrial stances are actually pretty mainstream these days. The Mexican eco-terrorists are just the most extreme case of an anti-humanist worldview that has been embraced very broadly across the green left and far beyond. Books and documentaries warning against progress, industry and even civilisation itself are a publishing sensation. We must renounce such hubris if we are to save the planet, we are repeatedly told not just by Mexican eco-anarchists, but by figures as thoroughly establishment as Prince Charles. Abandon the industrial in favour of the organic. Small is beautiful. Retreat to the rural, or at least the local. Restore the natural balance. Live simply. Raise chickens in your backyard.

Yet the left was not always this way. Historically, when we criticised the failings of the market, the left had no particular quarrel with industry, let alone science, technology or medicine. We celebrated modernity. Rather, our demand had always been that the fruits of civilisation should be extended to *all* of humanity. When did we turn away from the idea that capitalism was the problem, and begin to believe that it was modernity instead, or even the advent of mankind itself, that was the misfortune?

This book is my effort at answering the question as to why all this has come about. In many respects, it is merely a re-casting of traditional leftist arguments about industry, ambition and humanity's relationship with the rest of nature, but updated for the 21st Century—updated, you might say, for the Anthropocene.

It is a polemic, and at times an angry one. But the frustration comes not from a lack of worry about climate change or pollution or the difficulties thrown up by modern agriculture. It is precisely as a result of my concerns about these topics and my familiarity as a science writer with the reality of anthropogenic global warming that I am frustrated with what I argue are a series of romantic proposals from the green left that at best do very little to deal with the issue and at worst are counterproductive. Climate change is too grave a crisis to leave it to the greens.

At base, the book is a defence of industrial civilisation, scientific and technological progress, and economic growth. Each chapter counters one of a series of common arguments regarding limits to growth, carrying capacity, natural balance, anti-consumerism, local and organic food production, genetic modification, large-scale infrastructure, nuclear power and the notion of a 'metabolic rift' between urban and rural. But the book, particularly the final chapter, is also an attempt at intellectual history, exploring the counter-Enlightenment origins of anti-modernist green thought and the sharp divergence between such ideas and those of the left, or at least what the left historically stood for.

It is my hope that my arguments can assist in the revival of a pro-industrial, pro-growth left; that they might help galvanise those who are frustrated with the predominance of hair-shirted, anti-development greenery. I am convinced that our greatest hope in combatting a scale of climate change that significantly inhibits human flourishing lies in a turn away from a neoliberal emphasis on market-based mechanisms; ecological austerity; privatisation; localism; and regressive consumption taxes—but above all, away from Malthusianism, misanthropy and anti-modernism; and a turn toward a renewed enthusiasm for public-sector-led large-scale infrastructure; expansion of access to abundant, cheap energy; and an open-ended, steady raising of everyone's standard of living. Our best hope is for humans to keep getting happier, healthier, and yes, wealthier—but also more equal.

A renewed, modernist left is best placed to deliver this.

*

Ardent thanks are due above all to my parents, Lynda and Duncan, who offered me the space, time, quiet, encouragement, and a fridge sufficiently bursting with food and wine to enable me to finish my writing. The irony of drafting a book that celebrates industry, urbanity, and modernity while retreating to their home on the tiny, largely rural, crow-and-eagle-and-deer-supervised idyll of Bowen Island has not escaped me.

I also want to heartily thank my comrades and friends Eliyanna Kaiser, Ben Hayes, Oscar Reyes and Jason Walsh for their reviews of early drafts of the text and their thoughtful, expert suggestions. At the moment, I am quite *confident* about the ideas I have put forward in this book, but there is only one thing regarding them of which I am *certain*, and that is that there will be some positions in the coming years that I will not merely concede are incorrect, but will prove to be so mightily wrong that I will be acutely embarrassed to have believed them and likely try to deny that I ever did. I simply do not know which ideas they will be. I am grateful to the four of you for at least minimising this inevitable mortification.

But it is to Bruno Waterfield that I owe the most gratitude for nudging me in the right direction; for his clever, tight arguments over too many Westmalle Tripels and Lucky Strikes; for the sort of kind, mentoring friendship that everyone deserves but few are fortunate enough to find; for his model, bone-deep commitment to the dirty, scrappy, *noble* trade of journalism, without which there can be no democracy, no freedom.

1

The Apocalypse Is Bigger than Justin Bieber

When I was at university, during the apogee of what the media had dubbed the anti-globalisation movement[1] around the turn of the millennium, there was a group of campus activists that once a year celebrated something called 'Buy-Nothing Day'. The event took place on Black Friday, the first Friday after American Thanksgiving and the start of the Christmas shopping season. The idea was that everyone should protest our culture of over-consumption by boycotting *everything* for 24 hours. The activists aimed to 'raise awareness' of the environmental and social threat from unsustainable economic growth on the busiest shopping day of the year. 'We' should all consume less was the message.

It was first promoted in the late 90s by *Adbusters*, the Vancouver-based anti-consumerist magazine, and dovetailed perfectly with the publication of Naomi Klein's anti-branding and culture-jamming bestseller, *No Logo*, and the Black Bloc smashing of Starbucks and Niketown windows during the infamous Battle in Seattle mass protests against the World Trade Organisation.

But Buy-Nothing Day really irked me. I was one of those marching against the WTO ministerial meeting in the late November of 1999 and getting tear-gassed alongside the Teamsters and turtles. Together with a gaggle of eager young socialists, anarchists and environmentalists, faith community activists and trade unionists, I'd helped organize a series of coaches to come down to Washington state from British Columbia. Yet I was also a student maxing out on student loans and from a family that was really struggling at the time, having recently lost our home. So I was experiencing completely involuntary Buy-Nothing Days on a regular basis. I fervently wished for some 'Finally-Able-to-Buy-Lots-More Days'.

It was the assumption of equally grandiose levels of wealth in that little word 'we' in the demand that 'we all should consume less' that bothered me so much, the idea that 'we' in the West, every last one of us, were living a life of Riley, of carefree luxury and prosperity. I certainly didn't feel that I or many of my friends in similar situations were overconsuming at all. Instead, I felt like Tonto in that old *Mad Magazine* comic where he and the Lone Ranger are besieged by a throng of Indian braves, armed to the teeth. The Lone Ranger says to Tonto "What do we do, now?" And Tonto says back: "What you mean 'we,' kemosabe?"

The anti-consumerist, anti-growth argument has only extended itself since those heady street-fighting days. Great sections of the 'horizontalist' left have fallen under the sway of such deep-ecology thinkers as Derrick Jensen and Paul Kingsnorth who argue that industrial civilisation must be dismantled to varying degrees if we are to save the planet. The reality of climate change now requires that we overcome some of our most cherished ideas, says 'degrowth' guru Naomi Klein, writing in *The Nation* magazine: "These are profoundly challenging revelations for all of us raised on Enlightenment ideals of progress, unaccustomed to having our ambitions confined by natural boundaries."[2]

*

Writers with a jumble of doom-mongering-left and survivalist-right ideas can make a comfy living these days cranking out the Collapse Porn. In his peak-everything opus *The Long Emergency* and its anti-technology follow-up *Too Much Magic*, the *Atlantic Monthly*- and *Rolling Stone*-regular James Howard Kunstler seems to have a veritable hard-on for the end of the world, imagining with relish the coming Peak Oil collapse, a retreat from modernity and an embrace of the medieval. Just up the road from where I am writing this, the local toy store-cum-bookshop stocks Lego, Playmobil sets, Oprah-Winfrey-endorsed middlebrow fiction, Jamie Oliver cookbooks, and copies of *Unlearn, Rewild: Earth Skills, Ideas and Inspiration for the Future Primitive* by Miles Olson—a primitivist survival guide for the coming industrial apocalypse that teaches readers "radical self-reliance" skills such as animal hide-tanning, "feral food preparation," "natural methods of birth control" and how to shit in the woods. Above-ground poo-spot design requires no digging, Olson tells me. Covered with organic material such as leaves, moss and sticks, the poo pile can be used for quite a while, as it shrinks considerably as it decomposes. Dimitri Orlov, the Russian-American professional millenarian and author of *Reinventing Collapse* and *The Five Stages of Collapse* gives the book a hearty blurb endorsement: "It covers scavenging road kill, which grubs are good to eat and most everything else you need to know to go feral safely and in style." It's Scouting for Dystopians. I've only skimmed it, mind you, so I have no idea whether it also contains advice on how to deal with the sort of marauding gangs of hillbilly cannibals that you come across in Cormac McCarthy's *The Road*.

In the spring of 2014, *Vice* magazine—that ISIS of edgy, cross-platform media empires, built on a combination of fashion photography and hipster war-correspondency—dropped *Apocalypse Man*, a four-part documentary series profiling the former LAPD officer turned Peak-Oil-evangelist and author Michael Ruppert, who was already the subject of a critically regarded Armageddon-themed pop-documentary in 2009, *Collapse*. (Wrenching existential despair at the state of the world at least in part pushed Ruppert, also the host of *The Lifeboat Hour* radio programme, to shoot himself after recording his final broadcast of the show on 13 April, 2014.)

Surviving Progress, a 2011 eco-documentary executively produced by Martin Scorcese and featuring such figures as

Stephen Hawking and Margaret Atwood, makes the argument that past civilisations were destroyed by "'progress traps' – alluring technologies and belief systems that serve immediate needs, but ransom the future." Based on Canadian historian Ronald Wright's bestseller, *A Short History of Progress*, the film argues that today, once again, "progress is actually spiraling us downwards, towards collapse." Wright's pessimism in effect forecloses all possibility of ever improving humanity's lot: "Hope drives us to invent new fixes for old messes, which in turn create ever more dangerous messes." A similar conclusion is drawn in ecological historian John Michael Greer's *After Progress: Reason and Religion at the End of the Industrial Age*, attacking the "lab-coated high priests" of modernity (Greer seems to suffer a fondness for near-identical straplines: he is also the author of *The Long Descent: A User's Guide to the End of the Industrial Age*).

And the same argument is made by *End:Civ* — another of these insta-docs you can find tagged 'Controversial Documentaries' on Netflix sandwiched between *Super-size Me* and *Zeitgeist* and made by filmmakers filled to the brim with outrage but blithely indifferent to journalistic norms of evidence, fact-checking and careful, logical argumentation. Montages of household appliances, pig farms, instant popcorn and Caesar's Palace casino in Las Vegas are offered up as evidence of the wicked gluttony of civilisation, but without discussion; it is just assumed that the viewer will of course agree with the filmmaker, Franklin Lopez, that such phenomena are abominable, wasteful fripperies. Yet this documentary is laced with more radical conclusions than Wright's anti-humanist version of liberalism in *Surviving Progress* is willing to entertain, such as the need for eco-militants to blow up the Shasta hydroelectric dam in California; and indeed, *End:Civ* is based on deep ecologist Derrick Jensen's thesis in his own top-selling, two-volume call to arms, *Endgame* (*Volume I: The Problem of Civilization*; and *Volume II: Resistance*), that it is no less than the entirety of civilisation that is destroying life on the planet and so it all needs to come tumbling down. In the film, Jensen sermonizes:

> If civilization lasts another one or two hundred years, will the people then say of us, "Why did they not take it down?" Will they be as furious with us as I am with those who came before and stood by? I could very well hear those people who come after saying, "If they had taken it down, we would still have earthworms to feed the soil. We would have redwoods, and we would have oaks in California. We would still have frogs. We would still have other amphibians. I am starving because there are no salmon in the river, and you allowed the salmon to be killed so rich people could have cheap electricity for aluminium smelters. God damn you. God damn you all."

The film also features interviews with Jensen's fellow primitivist author Lierre Keith, the marine conservation Sea Shepherd Society's alpha-male-in-chief Paul Watson, and the aforementioned *Rolling Stone* anti-modernist scribe James Howard Kunstler. The documentary is a campus green-group film-night favourite across North America, while Jensen's books seem as mandatory an undergraduate bookshelf requirement these days as textbooks for first-year calculus or English Composition & Rhetoric.

The anti-consumerist, back-to-the-land, small-is-beautiful, civilisation-hating, progress-questioning ideology of degrowth, limits and retreat is hegemonic not just on the green left, but across the political spectrum. Far from being central to progressive thought, this cauldron of seething, effervescing misanthropy is in fact utterly alien to the rich tradition of humanism on the left and must be thoroughly excised from our ranks.

*

First, we must to ask these critics, if modern life is indeed rubbish, when was it exactly that humans enjoyed the 'correct' or sufficient standard of living? How much is okay and how much is now too much? When was it that we had it about right? Which was the period in the past when everything was copacetic?

Naomi Klein is perhaps the most high-profile of the de-growth, catastrophist and anti-consumerist thinkers. She is an award-winning author who has sold millions of books that have been translated into dozens of languages; regularly appears as a commentator in the media; frequently speaks at trade union conferences, environmentalist gatherings and left-wing teach-ins; has given talks at both TED and Occupy Wall Street; sits on the board of 350.org—one of the planet's most prominent climate change campaign groups; and was selected as the eleventh most influential public intellectual by *Prospect* and *Foreign Policy* magazines in their 2005 annual ranking of the world's Top 100 pointy heads.

And on the question of when precisely it was that humans lived in Edenic harmony with the Earth, Klein is all over the map. At one point, she is extraordinarily specific and the answer is surprisingly recent: it's the disco era. "The truth is that if we want to live within ecological limits, we would need to return to a lifestyle similar to the one we had in the 1970s, before consumption levels went crazy in the 1980s," she writes in her 2014 bestseller *This Changes Everything: Capitalism vs the Climate*.[3] At another point in the book, she says it was 1776[4] where we made the wrong turn, moving away from the natural rhythms of water wheels, which were suitably constrained by geography and "the flow and levels of rivers," to Watt's coal-fired steam engine, which radically released us from such limits, being deployable anywhere and at any time. But in an earlier essay in *The Nation*,[5] Klein identifies the Scientific Revolution as being our original sin:

Europeans—like indigenous people the world over—believed the planet to be a living organism, full of life-giving powers but also wrathful tempers. There were, for this reason, strong taboos against actions that would deform and desecrate "the mother," including mining. The metaphor changed with the unlocking of some (but by no means all) of nature's mysteries during the Scientific Revolution of the 1600s. With nature now cast as a machine, devoid of mystery or divinity, its component parts could be dammed, extracted and remade with impunity. Nature still sometimes appeared as a woman, but one easily dominated and subdued. In 1623 Sir Francis Bacon best encapsulated the new ethos when he wrote in *De Dignitate et Augmentis Scientiarum* that nature is to be "put in constraint, molded, and made as it were new by art and the hand of man."[6]

Elsewhere in *This Changes Everything*, Klein pushes the correct standard of living even further back to some indefinite but ancient Arcadia, prior to the advent of the Judeo-Christian[7] ideology of dominion over nature when we were more in touch with the land and less alienated from each other.

"The expansionist, extractive mindset, which has so long governed our relationship to nature, is what the climate crisis calls into question so fundamentally," Klein says. "We have pushed nature beyond its limits," she continues in her essay in *The Nation*, adding that the solution to not just climate change but all environmental problems is "a new civilizational paradigm, one grounded not in dominance over nature but in respect for natural cycles of renewal—and acutely sensitive to natural limits, including the limits of human intelligence."

James Howard Kunstler wants to hold down the civilisational rewind button longer than Klein, or at least when she's arguing that the 1970s was the optimum period after which everything went pear-shaped. In *The Long Emergency*, he proposes the Amish community as a model. Elsewhere he seems to favour a retreat to essentially a medieval level of development. "I'd propose that the whole world is apt to be going medieval," he writes in a piece decrying the popularity of French economist Thomas Picketty's book describing capitalism's inherent tendency toward inequality. Kunstler says Picketty deludes people into thinking we can still politically organise our way out of the civilisational predicament "as we contend with our energy predicament and its effects on wealth generation, banking, and all the other operations of modern capital. That is, they'll become a lot less modern." It is not merely that Kunstler believes a second Middle Ages is just another oil-shock away, but that he welcomes its arrival.

As does Paul Kingsnorth, the author of *Uncivilisation: The Dark Mountain Manifesto* and organizer of a series of similarly named festivals in the English or Scottish countryside celebrating his anti-modernist vision. *The New York Times* recently offered up a lengthy and largely flattering profile of Cumbria's high priest of primitivism and his neo-druidic eco-jamborees:

In the clearing, above a pyre, someone had erected a tall wicker sculpture in the shape of a tree, with dense gnarls and hanging hoops. Four men in masks knelt at the sculpture's base, at cardinal compass points. When midnight struck, a fifth man, his head shaved smooth and wearing a kimono, began to walk slowly around them. As he passed the masked figures, each ignited a yellow flare, until finally, his circuit complete, the bald man set the sculpture on fire. For a couple of minutes, it was quiet. Then as the wicker blazed, a soft chant passed through the crowd, the words only gradually becoming clear: "We are gathered. We are gathered. We are gathered."

...The hut was cramped and eerie, decorated with the

bones of small animals in illuminated glass cases. Haunting music was piped in from an iPod. You walked through a curtain, sat down and put on a heavy papier-mâché mask — a badger surrogate. Directly across from you, seated behind a window in the back wall, was another person — a volunteer — also wearing a badger mask. He or she sat silently, except when mirroring whatever movements you made, until, driven by emotion, fatigue, satisfaction or plain discomfort, you left.

Let's ignore for the moment the contradictions of opposing industrial civilisation while taking your iPod into the woods to provide a soundtrack to 'getting your feral on' and consider the key elements of Kingsnorth's anti-civilisational thesis. He is perhaps the most nihilistic of the current wave of popular writers aghast at what man has wrought, arguing that our current environmental problems are not even solvable. Rather, all that is left for us to do as collapse presents itself is to stoically endure (revel in?) our grief, despair and dread. For Kingsnorth, working through his eco-grieving appears to primarily involve giving workshops on the meditative benefits of using a traditional hand-wielded scythe as a low-carbon way to cut fields of hay, grain, grass and weeds.

In his manifesto, he proceeds with the customary doleful docket of (all too true) manmade ecological calamities that this sort of literature must apparently by contractual obligation regurgitate: a quarter of the world's mammals are threatened with imminent extinction; an acre and a half of rainforest is felled every second; 75% of the world's fish stocks are on the verge of collapse, etc. etc., alighting finally upon global warming:

And over it all looms runaway climate change. Climate change, which threatens to render all human projects irrelevant … which makes plain more effectively than any carefully constructed argument or optimistically defiant protest, how the machine's need for permanent growth will require us to destroy ourselves in its name. Climate change, which brings home at last our ultimate powerlessness.

There is nothing to be done. There is no economic or political reorganisation that will change matters; no 'techno-fix' that can rescue us at the last minute. Yet Kingsnorth's eloquent sarcasm (oh, and Kingsnorth is indeed eloquent; perhaps the most eloquent of misanthropes to have dipped his quill in bile since Schopenhauer's declaration, "Human existence must be a kind of error") at our ability to convince ourselves we can change course is not smug like Kunstler's, but merely rueful about our capacity for folly:

Daily we hear, too, of the many 'solutions' to these problems: solutions which usually involve the necessity of urgent political agreement and a judicious application of human technological genius. Things may be changing, runs the narrative, but there is nothing we cannot deal with here, folks. We perhaps need to move faster, more urgently. Certainly we need to accelerate the pace of research and development. We accept that we must become more 'sustainable'. But everything will be fine. There will still be growth, there will still be progress: these things will continue, because they have to continue, so they cannot do anything but continue.

Kingsnorth is the more literary twin of the main character in *Into the Wild*, the popular and poignant 2007 Hollywood film based on the true story of Christopher McCandless, the unfortunate young man who rejects what he feels are the hypocrisies of his suburban middle-class parents and their profit-driven mass society, gives away his savings to Oxfam, hitchhikes across North American and ultimately retreats to live off the land in the wilderness of Alaska, where he slowly dies of starvation. Where

McCandless had a film made about him by Sean Penn with a soundtrack by Eddie Vedder, Kingsnorth writes Booker-longlist-worthy novels about 11th Century Lincolnshire uprisings against Norman 'ingenga' (foreigners in Old English), the "fuccan swine in our own land," and similarly nationalist non-fiction laments for a departed England of village pubs and butchers' shops. (Tangentially, by the way—how different exactly is Kingsnorth's Anglo-Saxon nostalgia to the wistful Tory lament of John Major for his country of long shadows on cricket grounds and warm beer?) Underneath it all, Kingsnorth—like Klein, Jensen, the misguided McCandless and the rest—wants to "challenge the stories which underpin our civilization: the myth of progress, the myth of human centrality and the myth of separation from 'nature.'"

> Draw back the curtain, follow the tireless motion of cogs and wheels back to its source, and you will find the engine driving our civilisation: the myth of progress.
>
> The myth of progress is to us what the myth of god-given warrior prowess was to the Romans, or the myth of eternal salvation was to the conquistadors: without it, our efforts cannot be sustained. Onto the root stock of Western Christianity, the Enlightenment at its most optimistic grafted a vision of an Earthly paradise, towards which human effort guided by calculative reason could take us. Following this guidance, each generation will live a better life than the life of those that went before it. History becomes an escalator, and the only way is up.

Kingsnorth has a thing for the English Civil War. That seems to be the period to which he wants the clock turned back. Meanwhile anarcho-primitivist philosopher and Unabomber-confidant John Zerzan in his 1999 anthology of anti-modernist essays, *Against Civilization*, commonly stocked in your friendly neighbourhood lefty bookstore, sounds like a more writerly Alf Garnett grumbling about kids these days:

> It is impossible to scan a newspaper and miss the malignancy of daily life. See the multiple homicides, the 600-percent increase in teen suicide over the past 30 years [sic]; count the ways to be heavily drugged against reality ... Little wonder that myths, legends and folklore about gardens of Eden, Golden Ages, Elysian fields, lands of Cockaigne, and other primitivist paradises are a worldwide phenomenon, [a] longing for an aboriginal, unalienated state.

Meanwhile the author of the book's foreword, psychotherapist Chellis Glendinning, writes admiringly of the simplicity and superstition of Chicano village life in New Mexico:

> [A]n ancient and, until recently, undisturbed way of life. Men hunt elk and turkey. Women know plants. Curandera-healers with their potent prophetic powers live among us. Everyone knows how to build a mud house, dig the irrigation ditch, grow corn, ride a horse, and navigate through the forest on a moon-lit night ... there is more happiness here than in any place I have known. It's a simple happiness, nothing fancy.

Delving deeper still into this ideology, arriving at the maximum misanthropy terminus, the 'Deep Green Resistance' minions of Derrick Jensen meanwhile have fever dreams of a *total* collapse of human civilisation. Jensen would not stop at the Enlightenment and the Industrial Revolution. Like Zerzan, Jensen prefers a neo-lithic jam. For him, agriculture itself was the original sin: "the only sustainable level of technology is the Stone Age."

*

But, I hear you ask, indignantly clutching your signed, bright blue copy of *This Changes Everything*, is it really fair to place all these thinkers in the same degrowthist, anti-modernist basket (a basket that is presumably locally woven using only organic rattan and willow)? Isn't there something of a difference between the broadly social-democratic and liberal politics of the Guardian-or-New-York-Times-reading followers of a Naomi Klein or Bill McKibben that focuses upon civil disobedience and electoralism, and the anarcho-primitivism of the dreadlocked, soap-dodging myrmidons of a Derrick Jensen or John Zerzan, with its calls to insurrection and not infrequent terrorism apologetics? Or between the Anglo-mystic, scythe-bothering, mushroom-foraging, Wicker-Man aesthetic of a Paul Kingsnorth and the gonzo, shotguns-and-whiskey, profanity-and-spittle-spangled American off-grid individualism of a James Howard Kunstler? Klein and Jensen may have written pieces for Kingsnorth's beautiful hardbound journal, *Dark Mountain* (promoted on the apparently pre-industrial YouTube)[8]; and Kunstler may appear alongside Jensen in *End:Civ* (available on Pleistocene-era DVD from PM Press); and Zerzan, a sometime-comrade, sometime-rival of Jensen's, may have contributed to Ronald Wright's *Surviving Progress*; all of which suggests a rough cohort if not quite a school of anti-growth co-thinkers—but aren't there real ideological differences between them? Jensen, Kunstler and Kingsnorth appear to have given up all hope, while Klein feels that ecological collapse is not a foregone result, and that Kingsnorth-style grief can lead to change, arguing for things like public transport as a solution, not *Scything for Beginners* teach-ins.

At the most extreme end, some like Jensen make no apologies for the billions of human deaths that would accompany such a retreat; indeed, they welcome such a "die-off" as necessary. Jensen writes in the first volume of his 931-page doorstopper diptych of doom, *Endgame: The Problem of Civilization*, that he is even quite relaxed about no longer being able to access the

modern medicine he needs to deal with his Crohn's disease, that
he could be amongst the billions of humans to die:

> The truth is that I'm going to die someday, whether or not I
> stock up on pills. That's life. And if I die in the population
> reduction that takes place as a corrective to our having
> overshot carrying capacity, well, that's life too. Finally, if my
> death comes as part of something that serves the larger
> community, that helps stabilize and enrich the landbase of
> which I'm a part, so much the better.

Why concern ourselves with such manifestly human-hating,
even proto-genocidal philosophy? Surely Deep Green Resistance
eco-warrior militants advocating violence are on the margins of
eco-activism? Kingsnorth's retreats in the wilds of Dartmoor (or,
ahem, stately homes in Lanarkshire), with their smattering of
neo-paganism and a hint of English nationalism that pines for a
time before democracy, certainly do not attract the following that
the much more moderate eco-multinationals such as Greenpeace
and the Sierra Club do.

Certainly, yes, we can distinguish between these different
public intellectuals and their acolytes based on their tactics,
strategy, degree of pessimism, and how often they bathe. And we
can unquestionably cheer many of their campaigning efforts
against pollution, biodiversity loss and other very real ecological
challenges. My argument with them is not a critique of environmentalism *per se*. I myself write regularly about the science of
climate change. I march against Albertan tar sands. I am a consci-
entious recycler. I ride a bike.

Rather, what unites all these thinkers—and what I want to
contest—is the idea that we have gone too far, that *there are
natural limits to human flourishing beyond which we can never cross.*
Every last one of them believes there are fundamental boundaries to economic growth. Every last one of them is a haruspex

inspecting the entrails of industrial society and prophesying a frightful reckoning as a result of the hubris of man. Every last one of them lights a candle to venerate Icarus and Pandora, the patron saints of finger-waggers, doom-mongers and wet blankets.

Where they really differ is merely where these limits should be set. That is to say, which epoch they choose to have been the perfect time, when we "had enough": the 1970s, Amish Pennsylvania, the 1650s of the English Civil War, pre-Judeo-Christian civilisation, hunter-gatherer society, or some other time. Each author has their favourite era. But each of these prefer-ences is as *arbitrary* as any other. What makes one particular period superior to another appears more to be based on aesthetic affinity rather than any evidence of resource equilibrium between humans and their surroundings. Moreover, the reality is that thousands of young (and not so young) environmentally inclined activists are deeply influenced by these writers and the degrowth, anti-Enlightenment zeitgeist they have established.

And it's not just crusty anarcho-liberals in black hoodies with wonkily stitched-on 'ACAB' patches who argue that humanity has extended beyond the Earth's carrying capacity and that we need an economy that maintains a steady state or 'degrows'. Most environmental NGOs, progressive think-tanks from the New Economics Foundation to the Worldwatch Institute, Green parties and many left-wing and centre-left thinkers such as inter-national anti-carbon campaign group 350.org's Bill McKibben, and the former economics commissioner of the British Labour government's Sustainable Development Commission, Tim Jackson, take it as given that economic growth is the *central* problem to be overcome if we are to avoid climate chaos and even the end of life on Earth. In 2011, the late Ramon Fernandez Duran, a founder of *Ecologistas en Accion*—the 34,000-member Spanish environmentalist federation—and like Klein a leading figure in the anti-globalisation movement, published *The*

Breakdown of Global Capitalism: 2000–2030 – Preparing for the beginning of the collapse of industrial civilization. It is an unremittingly disconsolate, secular *Book of Revelation.* Duran writes at one point: "The endless progress of modernity is no longer a way to build heaven on earth and is becoming the construction of Gaia's hell. Hell has returned!"

There is also a growing international *'décroissance'* (French for 'degrowth') movement advocating for the contraction of production and consumption, centred on the writings of Serge Latouche, a professor of ecological economics at Paris-Sud 11 University and drawing on earlier work of Romanian economist Nicholas Georgescu-Roegen and well-known US ecologist and former senior World Bank economist Herman Daly. A largely Latin-centric front, the *décroissance* movement (*'decresimiento'* en espagnol; *'decresita'* en italiano) has had four conferences as of the time of writing, in Paris, Barcelona, Montreal and Venice. (More on their ideas later.) Traditionally, socialists have had no time for such arguments, but in recent years, a handful of leftist grandees such as John Bellamy Foster and David Harvey have come to concur that endless economic growth is not sustainable, even if they diverge somewhat from the conclusions of the more crunchy end of the degrowth spectrum.

*

Beyond green left campaigners, these ideas are firmly planted in mainstream consciousness. Klein's books, including *This Changes Everything*, are available from any high-street retailer. Walmart sells *The Derrick Jensen Reader* at a discount. In the *New Statesman*, pessimistic right-liberal philosopher John Gray gave his backing to the "stoical acceptance" of the insolubility of our environmental difficulties in Kingsnorth's *Uncivilisation*, while Roger Ebert described *Surviving Progress* as a "bone-chilling new documentary" worth three-and-a-half stars out of four that "tells the truth." A former director of the University of Toronto's Trudeau Centre for Peace and Conflict Studies, Thomas Homer-Dixon, published *The Upside of Down* in 2006, a similarly themed bit of apocalyptica predicting the breakdown of civilisation, but for the *Financial Times* set. Former war correspondent and liberal commentator Chris Hedges says "growth is the problem," as does leading environmentalist and member of the Order of Canada David Suzuki. Not one but two books titled *The End of Growth* have climbed the best-seller lists, one by former chief economist with CIBC World Markets Jeff Rubin, the other by journalist Richard Heinberg. The latter critic of growth is not only a senior fellow (alongside 350.org's Bill McKibben) with the California-based Post-Carbon Institute think-tank, but also an advisor to the king of Bhutan, and one of the series of lip-pursing 'thought leaders' to appear alongside former CIA director James Woolsey and top former-Soviet person Michael Gorbachev in *The 11th Hour*, yet another end-of-humanity documentary, this time from Leonardo diCaprio—the noted superyacht aficionado and UN Ambassador of Peace with a Special Focus on Climate Change. Heinberg seems to have made his career as a Peakist cassandra, also authoring *The Party's Over: Oil, War & the Fate of Industrial Societies; The Oil Depletion Protocol: A Plan to Avert Oil Wars, Terrorism & Economic Collapse;* and *Peak Everything: Waking Up to the Century of Declines.* Meanwhile, the aforementioned Post-Carbon Institute puts out reports with titles like *Climate After Growth: Why Environmentalists Must Embrace Post-growth Economics and Community Resilience,* and *Overdevelopment and the Delusion of Endless Growth.*

There are simply too many books to mention on this topic. English humorist Alan Coren was surely wrong when he said that the only books guaranteed to sell well were those about cats, golf and the Third Reich. His book *Golfing for Cats,* adorned with a Swastika, could also have done with some Peak-Oil pie-charts and a centerfold pull-out of a topless Martin Heidegger.

Meanwhile, establishment figures such as Sir David Attenborough and Prince Charles repeatedly admonish us all for our overconsumption and the madness of infinite growth on a finite planet. The former is also a patron of the anti-natalist Optimum Population Trust, recently renamed Population Matters, along with primatologist Jane Goodall, earth scientist and originator of the Gaia hypothesis James Lovelock, and former director of Friends of the Earth Sir Jonathan Porritt, as well as a glittering covey of other members, officers, commanders and knights grand cross of the Most Excellent Order of the British Empire and related orders of chivalry. The venerable Lester Brown, the octogenarian pioneer environmentalist and founder of the Earth Policy Institute, was hosted by the Smithsonian in 2012 at a symposium on the "unprecedented urgency" to act imposed by natural limits to growth. Geographer and WWF board member Jared Diamond warns us in his 2005 bestseller, also titled *Collapse*, that we are all headed the way of the extinct Easter Island civilisation as a result of overpopulation with respect to the carrying capacity of the planet. Writing in *Science*, the journal of the American Association for the Advancement of Science, Australia's former chief commissioner of its Climate Commission, Tim Flannery, called Diamond's tome "probably the most important book you will ever read."

And of course Hollywood pumps out eco-themed apocalypse blockbusters every summer, from *The Day After Tomorrow* to *Wall-E*, while an entire genre of film-making, the documentary, seems to have been all but colonised by collapse porn. The idea that we are living in the end times as a result of our consumptive greed and technological hubris—or as Slovenian philosopher Slavoj Zizek puts it in his book on the topic, we are "approaching an apocalyptic zero-point"—has taken over popular culture and consciousness. We see it in the viral popularity of articles such as the March 2014 *Guardian* article by Nafeez Ahmed breathlessly reporting that a NASA-funded study in just eight equations had mathematically proven industrial civilisation was on track to "irreversible collapse" (13,000+ shares on Facebook as of writing; 8,800+ tweets)[9]. We see it in that grating internet meme 'First World Problems', which assumes that all Westerners are equally, lavishly affluent these days and so the most harrowing experience we can undergo is having too much goat cheese in our salad, and that we have nothing in common with people in the developing world. And we see it in the hectoring craze for 'local' food and the rise of the Transition Town movement, eagerly readying itself for the eco-rapture by building backyard chicken coops, water wheels and sawdust-and-pathogen-encrusted compost toilets. There's a dedicated 'Collapse of Civilization' subreddit (http://reddit.com/r/collapse). Naomi Oreskes and Erik Conway, the great historians of late capitalism's phenomenon of scientists working as shills for Big Tobacco and Big Oil in their excellent, top-selling *Merchants of Doubt*, have recently published a new book: *The Collapse of Western Civilization: A view from the future*. Even the relatively sober *New Scientist* magazine in 2008 published a cover story entitled: "The Collapse of Civilization: It's more precarious than we realized," complete with a Stygian, sepulchre-hued painting of a city built on a precipice.

Derrick Jensen is fond of talking of his battle against "the dominant culture," a term taken from anthropology and sociology, believing himself to be a soldier in some sort of Gramscian counterhegemonic battle. Quite the contrary is the case. It is the counter-Enlightenment credo of that clutch of related concepts—degrowth, anti-consumerism, catastrophism, technophobia, localism and small-is-beautiful limits—that so dominates in contemporary culture.

It's time for progressives to remind themselves of the dark origins of anti-modernism, and understand that however well-meaning many of its supporters may be, this ideology is reactionary with respect to social progress and ultimately won't 'save the planet' anyway.

2

Austerity Ecology

The campaign against economic growth and overconsumption should have no place on the left. While its current austerity-ecology incarnation appears to many progressives as a fresh, new argument fit for the Anthropocene, it is in fact the descendent of a very old, dark and Malthusian set of ideas that the left historically did battle with. It is not that our species does not face profound environmental problems. Indeed, it is precisely *because* human society confronts such genuine ecological threats that the focus must be on the real systemic gremlins responsible for our predicament, not growth, let alone progress, industry or even civilisation itself.

Quite the opposite of all this misanthropy is what is imperative. There will need to be *more* growth, *more* progress, *more* industry and, above all, we will need to become *more* civilized, if we are to solve the global biocrisis.

To be clear, many of those on the green left who are concerned about the alleged problems of economic growth mean well, and for the most part, there should be no smug, sectarian derision directed their way. It's an absence of understanding of political economy that is at fault rather than conscious malevolence.

(Or at least that's what the po-faced and sensible little angel on my right shoulder tells me. The little devil on my left, a far more charismatic fellow at times, whispers instead: "Grant these hair-shirted GMO-free granola-druids no quarter. Remember that poster advertising a woodland gathering reading poetry 'to our brothers and sisters the trees'? I rest my case.")

*

Naomi Klein distils so much of this anti-humanist line of thinking into her 2014 bestseller, *This Changes Everything: Capitalism vs the Climate*, as well as in a handful of very widely shared essays appearing in *The Nation*, *The Guardian* and the *New Statesman*[10] from 2011 on that served as precursors to her recent opus. Let me note briefly that there is much that she says that I agree with. Above all, I doff my cap to her regular, robust promotion of trade unions and the rights of workers, something that too many other green-minded folks forget (most egregiously the world's most successful Green Party, Germany's *Die Grünen*, who together in government with the social democrats at the turn of the millennium broke the back of the country's union movement, laying the neo-mercantilist foundations of the current ongoing Eurozone crisis, a crisis in which the sizeable *Die Grünen* faction in the European Parliament has regularly backed EU policies that favour central European financial interests over those of the ordinary people of Greece, Spain, Portugal, Italy and Ireland). The single most important task for the left right now is a revival of the confidence of labour and a reversal of the working class's historic defeat by the forces of neoliberalism in the 1980s and 1990s. Nonetheless, due to Klein's prominence as a degrowthist thinker, and how representative she is of a much wider current, her arguments are going to figure prominently in this book's critique of anti-modernist ideology. Further, as we'll see, her degrowth arguments stand opposed to the interests of working people, and are a barrier to labour's advance.

In these texts, she puts forward the idea that climate change is actually something of a gift, a way for progressives to push through everything we've ever wanted but have never achieved. We can do this now because science tells us it's the only way to save the planet.

In her *Nation* essay, "Capitalism vs the Climate," she appears to make a revolutionary case against the market system after visiting a climate-sceptic conference hosted by the hard-right

Heartland Institute.

> If you ask the Heartlanders, climate change makes some kind of left-wing revolution virtually inevitable, which is precisely why they are so determined to deny its reality. Perhaps we should listen to their theories more closely—they might just understand something the left still doesn't get ... [C]limate change supercharges the pre-existing case for virtually every progressive demand on the books, binding them into a coherent agenda based on a clear scientific imperative.

She makes a similar argument in her *New Statesman* piece, "How science is telling us all to revolt." Here, Klein alights on the work of a pair of scientists with the Tyndall Centre—Britain's premier climate-research body—Kevin Anderson and Alice Bows, who have concluded: "We are now facing cuts so drastic that they challenge the fundamental logic of prioritising GDP growth above all else." Klein says that this in turn means that:

> [F]or any closet revolutionary who has ever dreamed of overthrowing the present economic order in favour of one a little less likely to cause Italian pensioners to hang themselves in their homes, this work should be of particular interest. Because it makes the ditching of that cruel system in favour of something new (and perhaps, with lots of work, better) no longer a matter of mere ideological preference but rather one of species-wide existential necessity.

A shortcut! All these past 200-plus years of systemic critique and political struggle from what we call the 'left', of campaigning, debating, voting, marching, picketing and on occasion revolting—in other words, the grand effort involved in putting forth our "mere ideological preference"—was insufficient because this was political rather than scientific. Now however, the men and women in lab coats have a secret weapon more effective than any boycotts, sit-ins, leafleting or electioneering; more certain to be victorious than any blockades, occupations or general strikes. All we have to do is present these facts and our ancient enemies will concede. Because all along, the problem in overcoming injustice has been that elites just didn't know the facts.

Okay, you say. So Klein is being a bit glib here about the magical power of The Science, (capital 'T', capital 'S') so what? And she's not saying don't engage in these other tactics of social change. Climate change is just an *additional* political opportunity. More importantly, beneath the hyperbole about the equations of atmospheric scientists and complex systems researchers telling us to man the barricades, isn't the substance of her argument the assertion that economic growth and overconsumption are behind what is causing climate change, indeed causing the wider biocrisis? Isn't she right that the Heartlanders are right?

Let's have a deeper look. Klein's argument involves two premises. The first is that to maintain ourselves within the range of average global temperatures that have been optimum for human flourishing throughout the Holocene geological epoch that began at the end of the last ice age, society must make radical and rapid cuts in carbon emissions, and ultimately move toward a carbon-neutral economy. Bows and Anderson put the scale of global cuts in CO_2 emissions needed after a peak in emissions in 2015 at a rate of 10 percent per year thereafter if we are to have an even chance of meeting the internationally agreed goal of a maximum two-degree temperature increase. Elsewhere, researchers with the Global Carbon Project put the figure at 5.5 percent per year over the next 45 years. For comparison, the fastest decarbonisation programmes in history, the transition to nuclear power by France, Sweden and Belgium in the 1970s and 1980s, with France now producing three quarters of its electricity from this low-carbon source, enjoyed reductions in emissions of four percent a year over the course of roughly a decade. Regardless of who is correct here, this is a *staggering* rate of carbon emissions mitigation. Klein is not wrong about the magnitude of the challenge.

Klein's second premise is that the main strategies favoured by policy makers up to now are at best inadequate and at worst counterproductive. She highlights in particular the failed strategies of green consumption, biofuels and carbon trading.

Again, so far, Klein is so lamentably accurate: whatever the benefits of driving a Toyota Prius and recycling your lentil tins, household consumption represents a much smaller percent of global CO_2 emissions than industrial emissions servicing business-to-business transactions. (To be fair, it is difficult to disaggregate these two parts of the economy. But to give a rough idea of the sort of ratio involved, Emilia Romagna in Italy, one of the wealthiest and most developed regions in Europe, assesses its ratio of household consumption to broader economic activities in terms of greenhouse gas emissions to be 27 percent to 73 percent[11].)

Meanwhile, first-generation biofuels have long been recognized as worse than fossil fuels once the emissions from indirect land-use change are taken into account. The sole biofuels that seem to be better than traditional fossil fuels when a full life-cycle analysis is performed are waste chip fat biodiesel (no really—this is a thing) and algae-based fuels. Unfortunately repurposing used chip fat is, to put it mildly, not very scalable, unless humanity started consuming bacon-wrapped jalapeño poppers, corn dogs and other deep-fried comestibles on a volume per capita basis in excess of that of a muumuu-wearing Homer Simpson. In the UK for example, current waste cooking oil supplies could power no more than one 350th of Britain's cars, according to the government's now-defunct Better Regulation Commission. And algae fuels at the moment would only be competitive with petrol if oil cost $800 a barrel.

And finally, the EU's flagship carbon reduction strategy, the Emissions Trading Scheme (ETS), is a cringeworthy disaster, with carbon credits dropping from around €30 ($41.50) a tonne in 2008 to under €3 a tonne by early 2013, with the European law enforcement agency Europol describing the system as an "open door" for organized crime[12]. Carbon prices have since picked up somewhat to around €5 a tonne at the time of writing, but this is still far below the €35 a tonne the designers of the scheme have always assumed would be necessary to spur market actors to make serious investments in low-carbon production technologies. Europe's slow but steady decline in carbon emissions since the advent of the ETS has been almost entirely a result of offshoring of production and drops in industrial output due to the economic crisis and nothing to do with carbon trading.

Klein concludes then on the basis of these two premises—that the required emissions reductions are colossal and the existing strategies for reductions aren't working—that the only remaining option with proven effectiveness is steep economic contraction, "what Anderson and Bows describe as 'radical and immediate de-growth strategies in the US, EU and other wealthy nations.'"

"Which is fine," Klein continues, "except that we happen to have an economic system that fetishises GDP growth above all else ... The bottom line is that an ecological crisis that has its roots in the overconsumption of natural resources must be addressed not just by improving the efficiency of our economies but by reducing the amount of material stuff we produce and consume."

De-growth and an end to overconsumption cannot be achieved without combatting capitalism, because capitalism is built upon these pillars—hence Klein's phrase, "Capitalism vs the Climate." It does look at first glance as though revolution becomes, as she puts it, a species-wide existential necessity.

*

The first point I really want to underscore here is that one cannot in one breath rage against the imposition of economic austerity — the series of radical cuts to social programmes and depression of wages imposed by most Western governments in the wake of the global economic crisis — while arguing against economic growth. Austerity and 'degrowth' are mathematically and socially identical. They are the same thing. What green degrowth partisans are actually calling for is eco-austerity.

This is because if your starting point is that humans are consuming too much, then *any* cuts to social programmes and wages will result in less money in these same humans' pockets, and hence less consumption. So however cruel austerity may appear, you really should be cheering this on. And yet Klein elsewhere sharply and correctly criticises the injustice of austerity. Likewise, we frequently we see the same people marching against cuts to social programmes and for an increase in wages — whether as part of Occupy Wall Street-type action in New York, or the Fight-for-15 campaign of fast-food workers for a living wage in Chicago and 150 other cities across the US, or in more street-fighting fashion against the fiats of the Eurozone on the streets of Athens or Barcelona — who, come Earth Day or some climate-change protest, will raise different placards, this time damning economic growth. But the two positions are irreconcilable.

Unlike some anti-growth proponents, Klein does at least concede that the only historic comparison we have to a 10 percent drop in emissions year after year is that of the economic contraction during the Great Depression. Emissions reductions after the 2008 crash averaged only seven percent across the OECD, and only for one year before rebounding, and the Soviet Union saw reductions of five percent over ten years. To avoid such social catastrophe, Klein says that the economic contraction must be carefully managed. Nevertheless, Klein repeatedly argues that "we all" need to consume less, just like the Buy

Nothing Day activists and thousands of other green campaigners. What is the most famous green slogan, but *"Reduce, Re-use, Recycle"*? Klein and others—regardless of how they couch their calls to degrowth, no matter if the emphasis for now is on reducing the consumption of the wealthiest first, and only later restricting the consumption of the rest of us—are still saying that even the most equitably managed contraction would involve *a reduction in every Westerner's standard of living.*

At this point, it is worthwhile revisiting Klein's choice of the 1970s as one of her favoured eras of eco-arcadia, after which, as she told UK writer Owen Jones in a public talk in London, "we" (everywhere this blasted undifferentiating first-person plural pronoun) "turbo-charged the American Dream in the eighties" and, as she writes in the book, in the 1990s, proceeded to "take it global." From such phrases, one could be forgiven for thinking that the high rate of expansion of the standard of living of ordinary people that occurred from the late 1940s through to the mid-1970s in most Western countries continued apace into the 1980s and up till today. But this is false. In fact, the opposite is the case. There is a reason the French call this post-war epoch *Les Trente Glorieuses*—The Thirty Glorious Years of high productivity, high wages, full employment, expanding social benefits, powerful trade unions and increasing consumption—and not *Les Soixante-dix Glorieuses* as it surely would be if Klein's periodisation obtained. By almost every measure you could come up with, the economic standing of ordinary people has stagnated or declined since the late 1970s.

In the late 1960s, the Keynesian mixed economy began to sputter amid rising inflation, diminished profits and a crisis of rising expectations on the part of workers—in other words an increasing trade union militancy that was the natural result of policies of full employment, predicted in Polish economist Michal Kalecki's famous 1943 essay, "The Political Aspects of Full Employment":

[U]nder a regime of permanent full employment, the 'sack' would cease to play its role as a 'disciplinary measure'. The social position of the boss would be undermined, and the self-assurance and class-consciousness of the working class would grow. Strikes for wage increases and improvements in conditions of work would create political tension.

(Put very simply: If you can get a job tomorrow somewhere else, why put up with low wages or a domineering boss?)

As a result, the Keynesian consensus broke down, to be replaced with a basket of inegalitarian policies building on neoclassical economics that today are described as 'neoliberalism'. This austerity was initially sharply resisted by working people and the trade unions throughout the 1970s and early 1980s, but by the end of that decade, the union militancy of the 1970s had largely been smashed and full employment was a distant memory. Although most associated with right-wingers such as Margaret Thatcher in the UK, Ronald Reagan in the US and the Chilean dictator General Augusto Pinochet, neoliberalism has been imposed with similar élan by social democratic governments, trapped as they are by an ideology that has no systemic critique and so restricts the horizon of the possible to more fairly sharing out the spoils of capitalism amid periods of growth and more fairly sharing out the pain amid periods of stagnation or crisis.

Then, to continue with our potted, far-too-brief economic history of the last few decades, in 1991, the Soviet Union collapsed. The USSR was a savage, peasant-consuming Polyphemus of a regime whose demise I unreservedly celebrate. No one of the least progressive bent should have any time for the useful idiots of Stalin or Mao who continue in surprising numbers to this day to apologise for Communist wickedness. Those cool kids sporting 'Full Communism Now' t-shirts and clasping the latest publication from whoever is the Stalinism-minimising obscurantist Continental philosopher of the month, deploying the scandalous c-word *pour épater le bourgeois,* can fuck off. Nevertheless, we must be frank and recognize at the same time that until its demise, the USSR served as an outsized, bowel-loosening foghorn of a warning to Western elites that immiseration of ordinary people would likely result in their own overthrow, and a subtle reminder to those very ordinary people that while Stalinism may not be the preferred alternative to capitalism, alternatives to capitalism at least existed. So with the Soviet Union gone, both the masters of the universe and their subjects were now convinced that there is no alternative. It was the end of history.

While productivity in the US grew 80.4 percent between 1973 and 2011, according to the centre-left Economic Policy Institute, median worker pay grew just 10.7 percent. Where during *Les Trente Glorieuses,* workers' pay rose in tandem with productivity, since the 1970s, there have been only stagnant or declining wages and benefits. Whichever metrics we use, we arrive at the same result. Working families in American make less than they did 15 years ago—a phenomenon that has not occurred since the Great Depresssion. Since 2000, weekly earnings for low- and middle-income workers have remained essentially unchanged, after adjustment for inflation. If the median US household income had kept pace with the broader economy since 1970, it would now be $92,000 instead of $50,000. And of course, similar numbers can be found for other jurisdictions. From 1990 to 2009, labour's share of national income declined in 26 out of 30 developed economies for which this data is available, according to the OECD. Overall across the advanced economies, labour's share dropped from 66.1 percent to 61.7 percent. Meanwhile the depth of this decline in developing countries is even more pronounced, according to the ILO, with steep falls in Asia and North Africa and stable but still declining shares in Latin America. It's even happening in China.

So Klein's call to roll back "our" standard of living to the 1970s is simply, egregiously, ahistorical. The truth is that for most people, we never really left the seventies.

It should be noted here that some radical green activists such as Derrick Jensen do however recognize this dissonance between calling for decreased consumption and opposing austerity. Going further than Klein is willing to, they do not shy away from actually embracing economic crisis and its accompanying social fall-out, or they complain that any time union members fight for higher wages, they are mounting a defence of their privilege and waging war against the planet because they will now be able to consume more. Openly favouring other organisms over humans, they argue that a reduction in living standards, contra the historic position of trade unions, is simply the price that must be paid by our one species for the sake of the rest of the biosphere. In a 2000 talk to a convention of the Green Party of Canada, the late Canadian theoretician of the 'left biocentrism' wing of deep-ecology David Orton, echoing the ideas of Rudolf Bahro, a founder and federal executive member of the German Greens, said that he "sees the trade unions as united with their employers in defending industrial society and privilege ... Both unions and employers have an economic interest in the continuation of industrial society and speak with similar anti-ecological voices. In the main, of course there are exceptions, trade unions are generally environmental enemies, not allies, of the environmental and green movements."

Such "checking the privilege" of trade unions must be music to the ears of employers.

*

This new paradigm of rejecting growth and embracing limits is also by definition a rejection of progress. It is to say: this much and no more. Or, more precisely, that we can expand but only in *non-material* forms. Klein, for example, emphasises that her prescription is "selective degrowth," which she clarifies in a 2014 interview with the New York *Indypendent* newspaper: "There are parts of our economy that we want to expand that have a minimal environmental impact, such as the care-giving professions, education, the arts. Expanding those sectors creates jobs, well-being and more equal societies." But the *material* side of the economy—the "extractivist" side, in Klein's words—has to shrink.

All this voluntary-simplicity, simple-living rhetoric sounds lovely, warm and fuzzy. I'm certainly feeling the feels when I read plaintive yearnings in popular environmentalist magazines like *Orion* or *Grist* about building community, or overhear the kale-wranglers and turnip-whisperers at my local farmers' market pining for a society where we are more neighbourly and devote more time to friends and family, art, poetry and music. But all this sort of "embracing other, less material ways of well-being" ignores that you can't make music without instruments or write poetry without ink and paper, and instruments and paper can't be made without raw materials that need to be chopped down or mined. A whistle is made of tin and a trumpet made of brass. This argument (or mood, really; it's less an argument than a sentiment) also forgets that it is increased productivity through technological advance (combined with trade union organising) that gives us more free time that would allow us to be more neighbourly and community-oriented. So this immateriality of "other kinds of growth," of "selective degrowth," is a fantasy. While we *can* steadily dematerialise production via technological innovation, and though knowledge itself is certainly immaterial, knowledge will always be *linked* to the material, both in its origins and its products. New knowledge depends on old technologies, old *stuff*, and gives rise to new technologies, to new *stuff*.

Think about it this way: if we have retreated to the optimum economic stasis-point of the Kleinian imaginary, where we are supposed to no longer be overshooting our carrying capacity, then each one of us has all the right amount of 'stuff'—no more and no less. But now, if through the expansion of our knowledge, we develop a new technology that does not replace—or only partly replaces—a previous technology, and yet we want to put it into production because of its manifest benefits to society, then we will have to give up production of some other technology to make room for it. But hold on—we've already decided that we have all the stuff that we need, no more and no less. That means that we cannot give up that old technology. Thus we either invent nothing new (or at least only those new technologies that perfectly replace old technologies without any overall expansion of production), or we have to grow. Therefore, the steady-state economy must by definition refuse most technological advance, and even most new knowledge as well. The steady-state economy is a steady-technology economy, a steady-science economy. It is a static society, the very definition of conservatism.

*

Since Karl Marx in 1858 tartly described the Reverend Thomas Robert Malthus as "this baboon" for his 1798 work, *An Essay on the Principle of Population*, which had made some of the very first overconsumption arguments, the left has mounted a fierce battle with these ideas, arguing that for all their seeming society-wide concern, at best they obfuscated the systemic source of injustice and at worst exacerbated capitalist knavery.

But for some reason, the idea of overconsumption in recent years has been getting a warmer welcome in certain leftish quarters, even amongst those who continue to battle the overpopulation discourse. Such a contradictory position is as tenable as a Temperance League gin-tasting fundraiser.

Ecosocialists Ian Angus and Simon Butler in 2011 published *Too Many People?*, in the main a tremendous volume carefully pulling apart all the false arguments of the advocates of the overpopulation thesis, but only to conclude that economic growth however *is* a problem! Yet they are the self-same discussion. Either you need to restrict the number of people, given a certain amount of resources, or you need to restrict the amount of resources used, given a certain number of people.

Put another way, if we had the egalitarian system that these leftists (and I) want, and if population continued to grow but economic output did not, then there would be an ever-diminishing standard of living as the per capita output shrank. Economic growth must keep pace with population growth. So either the Malthusians are correct and overpopulation is a problem, or overpopulation is not a problem, but then neither are overconsumption and growth.

Even Jensen recognises the identity of the two arguments: "There are simply too many people. You've seen the pictures. Crowded streets in Calcutta, impoverished babies with huge hungry eyes and bloated bellies in Mexico, refugee camps in Africa, masses of Chinese crammed into filthy cities. The earth can't support these numbers. Something's got to give ... Another way to talk about this is to notice the language: overpopulation, zero population growth. How different would our discourse be if we spoke instead of overconsumption and zero consumption growth? This shift in discourse won't happen, of course, because zero consumption growth would destroy the capitalist economy."

Relatedly, for all his deep misanthropy, the primitivist Jensen with his wish for us to return to hunter-gatherer society is at least consistent here compared to zero-growth advocates that pick some other era. If progress and growth are the problem, then we must return to a time when there was no growth or progress, not pick some random period due to aesthetic affinity. There was still progress and growth prior to the Industrial Revolution, and before the Enlightenment, and before the Renaissance, and in the Middle Ages or Tang Dynasty China, and in Ancient Rome and in

Mesopotamia. Of all the flavours of degrowthism, only Jensen and Zerzan come the closest to consistency when they look to primitive society, noting that for the bulk of our existence, we were pretty much a zero-growth society: the roughly 178,000 years between the emergence of anatomically modern *homo sapiens* in Africa some 200,000 years ago, and the rise of sedentary agriculture, also known as the Neolithic Revolution, that began an estimated 12,000 years ago in the Fertile Crescent.

But even here, anthropologists do not declare this epoch to be *absolutely* zero-growth, only that growth was so slow as to be imperceptible to those living through it. The shift from an animal ethology (animal behaviour) to behavioural modernity around 50,000 years ago is the site of great controversy amongst researchers. It's a bit of a slippery term, but behavioural modernity describes the point at which our species, *homo sapiens*, began to employ complex symbolic thought. In other words, language, abstract reasoning, advanced problem solving and artistic symbolism—culture; as well as the ability to invent, rather than merely adopt, new practices. The dispute in the scientific community is whether this arose via a technological revolution in the upper Paleolithic over the course of 10,000 years, or, as partisans of the 'continuity hypothesis' suggest, much more gradually over hundreds of thousands of years. But regardless of the true time period, no one should confuse either with stasis.

As for Jensen's favouring of this epoch for it being the last time that humans lived in balance with the rest of nature, never taking too much from the land-base, one has to of course confront the awkward reality of the extinction of Pleistocene megafauna—mastadons, dire wolves, giant condors with wing-spans of 16 feet, nine-foot-long saber-toothed salmon—every-where that Paleolithic humans went, a process known as the Quaternary extinction event. This in turn comprehensively altered terrestrial ecosystems further through cascading changes to the food web when apex predators were removed, involving a reduction in lateral diffusion of mineral nutrients, and perhaps, due to the disappearance of megaherbivores such as woolly mammoths and bison, a sharp reduction in atmospheric methane, which in turn may be implicated in the sudden climatic cooling that lasted for about 1300 years known as the Younger Dryas or Big Freeze. Caveman climate change?

Jensen responds to this by simply denying that this happened, as it does not fit in with his belief system that holds indigenous peoples to have lived a sylvan existence in perfect harmony with the environment, quoting the work of one author, ethnobiologist Eugene Hunn. While it is true that there remains an open debate about this period, there is increasingly wide consensus amongst paleobiologists and ecologists that humans played a major role, either through direct overkill or through out-competing other predators, or in some interplay of various factors. Jensen's dependence on just one source here to deny anthropogenic pleistocene extinction parallels the single-study syndrome of climate sceptics who pounce on one new study in an obscure journal purporting to show that sea-levels aren't rising in order to deny anthropogenic global warming.

It should go without saying that it is more than possible to decry the genocide of the indigenous peoples of the Americas, Australasia and Africa, and to campaign against the wretched racism they continue to face, without having to embrace the ideal of the Noble Savage, which is racist in its own, ethnically essentialising way. American Indian ethnohistorian David Rich Lewis gives a convincing history of the invention of the modern equivalent of the Noble Savage, the trope of 'Ecological Indian', tracing it through 60s counterculture, the Crying Indian of the 1970s' 'Keep America Beautiful' advertising campaigns, a script writer's creation of a fictional speech by Chief Seattle, right up to Disney's *Pocahontas* cartoon and Kevin Costner's white-saviour epic, *Dances With Wolves*.

People searching for a new relationship with nature and a set of spiritual values to counter the individualism, political economy and environmental impact of modern industrial society latched on to the image of the Ecological Indian. Ideals of Indian communalism and ecological relationships rooted in a spiritual harmony with Mother Earth attracted those looking for alternatives ... [T]he ancient belief in a Mother Earth was more the creation of modern scholars looking for religious universals than a pervasive pan-Indian concept, but if filled important rhetorical and emotional needs. Environmental organisations joined native groups in arguing for Indian rights and an Indian-centred environmental model, and Indians in turn internalized the rhetoric of Mother Earth and the Ecological Indian for their own political and intellectual purposes.[13]

Lewis goes on to recount the scholarly consensus that far from encountering a pristine wilderness when Columbus arrived in the New World, he and his fellow butchers that came after found extensively anthropogenic landscapes. Native peoples locally overhunted a number of extant species such as seal, sea lion, musk oxen and caribou; as aquaculturalists were "deadly efficient" in their construction of weirs and dams; and created vast trails and roads serving extensive trade networks. They built cities, terraced hillsides and irrigated fields, shaped the land with intentional fires, and cleared forests, all of which contributed directly to deforestation, soil depletion and erosion. In forgetting all these works, the myth of the Ecological Indian infantilises and denies agency to native peoples, erases civilisations and replaces them with imaginary wildernesses. It too is an act of conquest.

*

But regardless of contemporary oversimplifications of the tale of flawless indigenous resource management, if the advent of agriculture was the original sin, as Jensen and many others such as the much more moderate Jared Diamond argue, it was a sin that humans in many different places were keen to engage in over thousands of years, as there have been eight to ten different geographic sites with contrasting ecosystems that have been identified by researchers as probable independent locations for the origin of domestication and agriculture, each with distinct sequences, species and methods. The precise causes of the shift from hunting and gathering to agriculture and settlement remains obscure, but all developed though whenever and wherever clever, clever humans experienced local variations on the following thought process that must have happened in the southern Levant in the Near East some 11,400 years ago: "Aha! If I cut and replant branches of these fig trees here that produce sweeter, softer fruits, next year, we'll have more yummy figs and fewer yucky figs!"

The innovation of agriculture, aided by our species' uniquely prodigious capacity for problem-solving and ingenuity, was perhaps driven by the loss of calorific abundance following our driving so many megafaunal beasts to extinction (a very early supersession of Malthusian 'limits to growth'). Or perhaps it was simply a slow but steady, or clunky and herky-jerky, non-linear accumulation of small innovations.

But regardless of the contours of its invention, perhaps Jensen would respond to the advent of such agronomic wickedness: did we really *need* those figs of a finer flavour? Surely our core homeostatic needs of warmth, food, water and oxygen were already largely being met. This was a luxury item, not a need!

Thus we will have to banish all innovation, imagination, intelligence and desire for slightly sweeter figs (or barley or flax or bitter vetch or whatever) if we're going to be able to ban growth. In which case, progress begins tens of thousands of years earlier than agriculture, with the invention of spears, needles, string, nets, harpoons, snares and fishing tackle. It's a remarkable thing, but canoes, sailing, knowledge of ocean currents, and even celestial navigation pre-date agriculture. We are marvellous creatures. We've been doing nothing but innovating for tens of thousands of years, perhaps much longer than that. The controlled use of fire pre-dates the existence of our species, likely a discovery of our ancestor, *homo erectus*, 400,000 years ago.

So if Jensen and Zerzan and their army of money-wrenching minions were somehow successful in achieving their Glorious Primitivist Revolution, they would have to be ever on their guard against cavemen boffins getting any bright ideas about how to make life easier or more pleasurable. It's all a rather Pol-Pot-ian Year-Zero-type regime they want to inaugurate, no? Or perhaps just an innovation Stasi? (Perhaps this is too ghoulish a joke though. The Khmer Rouge did actually implement Egyptian-French Maoist economist Samir Amin's advice of "rapid disurbanisation" and autarky. Amin continued to endorse the Cambodian regime as late as 1981, before retreating from such a stance, as they had embraced precisely the policies of radically cutting themselves off from the world economy and "self-sufficiency" in agriculture ['food security', in contemporary parlance perhaps?] that he recommended be adopted across Africa. Likewise, when news emerged of intellectuals being deported to the countryside, Malcolm Caldwell, the then-chair of the Campaign for Nuclear Disarmament and the most prominent Western apologist for 'Democratic Kampuchea', wrote: "No doubt it will be hard for some urban dwellers accustomed to pushing pens or turning ledgers to adjust to the labour in rice-fields, but such hardships as may arise cannot be construed as a bloodbath, unless many commit suicide rather than submit to it."[14] However insane Jensen's programme for change may seem, we shouldn't laugh, as it has been put into genocidal practice before, vigorously and callously defended by leftists such as Caldwell, who himself was ultimately shot in mysterious circumstances during a visit to Phnom Penh.)

Of course, one might argue that I'm being far too loose with the terms growth, progress, and invention, which begin to blur here. But then, as well they should, as perhaps what it means to be human is to invent, to progress, to grow. To constantly strive for an improvement in our condition. To overcome all barriers in our way. To never at any point sit back and relax and say: "Ah, I think we're done here. We have enough. Life does not need to get any better. These bitter figs will do."

*

To clarify the contemporary anti-modernist discourse and understand how we have arrived in this intellectual cul-de-sac, perhaps it is worthwhile to revisit some of the ideas of Malthus and his epigones directly. The cleric postulated that left unchecked, population grows geometrically (today, we would say exponentially: two people, then four people, then eight, then 16, 32, 64, 128, and so on) while the amount of food we can produce grows only arithmetically (today, we would say linearly: two sacks of wheat, then three sacks of wheat, then four, then five, six, seven, and so on). Humanity was no different from any other life-form in this regard, and so just as "the constant tendency in all animated life to increase beyond the nourishment prepared for it" results in its being checked by famine and disease, so will the human population. He fervently opposed relief for the poor, as this only encouraged the lower orders to breed, hastening the looming catastrophe, when instead "unwholesome occupations," vice, poverty and disease should be left to work naturally as a check on population growth:

> [W]e should facilitate, instead of foolishly and vainly endeav-
> ouring to impede, the operations of nature in producing this

mortality; and if we dread the too frequent visitation of the horrid form of famine, we should sedulously encourage the other forms of destruction, which we compel nature to use. Instead of recommending cleanliness to the poor, we should encourage contrary habits. In our towns we should make the streets narrower, crowd more people into the houses, and court the return of the plague.

Without such measures, he says, disaster awaits:

Should success be still incomplete, gigantic inevitable famine stalks in the rear, and with one mighty blow levels the population with the food of the world.

But the predicted Malthusian catastrophe never materialized. Extraordinary productivity increases during the Second Agricultural Revolution resulting from improved crop rotation, better plows, selective breeding, the seed drill and new drainage techniques—but also, we cannot forget, the enclosure of the commons and the rise of capitalist farmers—saw food supplies rise faster than population and a metamorphosis from feudal subsistence to an early market economy. This agrarian transformation revolutionised labour as well, releasing agricultural workers to be employed in the dark Satanic mills of the Industrial Revolution.

Much later, the Green Revolution led by US biologist Norman Borlaug in the 1940s and 50s transformed agriculture via modern plant-breeding, irrigation infrastructure, synthetic fertilisers and pesticides. Scientists developed hybrid rice and wheat varieties that responded better to plant nutrients and grew with shorter, stiffer stalks to support the heavier heads of grain that produced higher yields. Agricultural productivity exploded in this period, simultaneously expanding the food supply and slashing prices. (Contemporary critics of the Green Revolution have some legitimate concerns about how these advances led to a preference for monocultural production and a shift from agricultural inputs being generated on-farm to the purchase of inputs, in turn producing greater farmer debtloads, the loss of their land and ultimately the concentration of land ownership in fewer and fewer hands. But all this is should be viewed as a criticism of an inegalitarian market-based mode of production rather than of the technological advances.)

This remarkable achievement of the Green Revolution was only met with more catastrophism by a new generation of Malthusians. *Famine-1975! America's Decision: Who Will Survive?* was a 1967 data-heavy bestseller by plant pathologist William Paddock, who himself had developed a disease-resistant strain of corn high in Vitamin A, and his diplomat brother, Paul. The book is long-forgotten and a bargain-bin filler at used bookshops today. But *The Population Bomb* by entomologist Paul Ehrlich, arguing like the Malthus that the Green Revolution had only encouraged the damned rabbits to produce more of themselves, made a much bigger splash. He predicted in 1968, according to his extrapolations, that an acceleration in the world's population was certain to produce global food shortages that would result in as many as four billion dead over the course of the 1980s. Also like the dear reverend's call to end poor relief, Ehrlich wanted to see an end to development aid, but even this would probably be insufficient. "The battle to feed all of humanity is over. In the 1970s the world will undergo famines – hundreds of millions of people will starve to death in spite of any crash programmes embarked upon now." A latter-day Malthus in many ways, he advocated cuts to public funding of healthcare ("death control"), making foreign aid conditional to adopting population control measures, the sterilisation of all fathers of three or more children in the developing world, and even considered putting sterilising chemicals in municipal water supplies and punitive taxes on diapers and cribs.[15]

Ehrlich was invited onto the Tonight Show some 20 times, making population growth a hot topic amongst the American public. The book prompted the establishment of the first Earth Day in 1970 as overpopulation was believed to be the cause of all other environmental problems (a concern echoed on Earth Day in 2014 when the Center for Biological Diversity handed out 350,000 Endangered Species Condoms across the US to remind attendees of "runaway human population growth and overconsumption"). Richard Nixon embraced the topic with gusto, declaring the population crisis to be "one of the most serious challenges to human destiny in the last third of this century," while the US Congress established a Commission on Population Growth and the American Future.

What happened? All other things being equal, Ehrlich's models should have been right. But all other things didn't stay equal.

Firstly, the techniques of the Green Revolution were spread still further afield. But much more importantly, by the end of the 20th Century, almost all developed countries had seen a sharp drop in fertility rates as the result of the spread of effective birth control, but also as a result of economic development (growth!), education and the lower infant mortality that came with better healthcare. According to the predictions, lower infant mortality was supposed to *add* to the population burden. Instead, as families now could be pretty assured that all their offspring would survive childhood, there was no need to have as many children. Contra Malthus, who believed that an improved standard of living, greater means, would result in a larger number of offspring, the opposite is quite demonstrably the case. Such transformations were beginning to be apparent precisely during the high point of the 70s population scare. While Ehrlich was chatting with Johnny Carson, Karen Singh, India's then population minister, was already declaring: "Development is the best contraceptive."

Today, there is almost no controversy amongst demographers about this phenomenon, known as the 'demographic-economic paradox': the higher per capita GDP of a population group, the fewer children born, even if every few years Chicken Little Ehrlich sounds the alarm once again with fresh extrapolations to prove that this time, he's right. As Angus and Butler point out, we now know that world population growth rates have been slowing down since their peak in 1963 — five years *before* the publication of *The Population Bomb*. The total fertility rate — that is, the number of children a woman will have on average over the course of her lifetime — is now below replacement level in more than 116 countries.

We are at the point that some demographers are concerned that the EU's population could drop by a quarter and Japan's by a half within 45 years. The UN Population Database low-variant projection puts global population topping out around mid-century with just over 8 billion people and dropping to around 6 billion by century's end, although the high variant projection assumes continued growth to 16.6 billion by 2100. Angus and Butler quote demographer and mathematical biologist Joel Cohen: "In spite of the abundant data to the contrary, many people believe that the human population grows exponentially. It probably never has and probably never will."

(One might add tangentially here that for decades, computer scientists have been hard at work trying to develop artificial general intelligence supercomputers [or 'strong AI'] that will be able to do incredible, amazing, stupendous things. But already we have 7 billion absolutely incredible supercomputers just walking around farting, fucking, writing symphonies and designing gravitational wave telescopes to peer back in time to the earliest, tiniest fractions of a second of the infant universe. And more humans means more awesomeness. More novels, more jazz, more medical advances, more refrigerator-sized spacecraft landing on comets 515 fucking million kilometres from Earth.

Why would we want fewer of these incredible, amazing, stupendous *meatware* supercomputers anyway?)

So Ehrlich and his co-thinkers were laughably wrong, yet, unlike the contemporary consensus on anthropogenic climate change, there has *never* been any consensus in the scientific community about overpopulation, but this has not prevented organisations such as the Sierra Club, the World Wildlife Fund, the Royal Swedish Academy of Sciences and the UN from showering Ehrlich with awards. Ehrlich, a former adviser to the board of directors of the anti-immigration Federation of Americans for Immigration Reform (indeed, as we will see, overpopulation politics very frequently overlaps with anti-immigrant politics), was also made a fellow of the Royal Society in 2012. All these gongs and laurels are probably why Ehrlich has not felt any need to refrain from every few years just shifting the date for the end of the world a little further.

Indeed, the long-standing establishment embrace of the Malthusian worry explodes any suggestion that anti-growth is a particularly radical position. But then elites have never really liked the vast bulk of humanity, believing us all to be lesser creatures, unlettered, unrefined.

Four years after Ehrlich published his seminal human-hating tract, the Club of Rome think-tank, whose membership is composed of former heads of state, senior UN civil servants and business leaders—a more establishment body one could not find—produced its own report funded by the Volkswagen Foundation called *Limits to Growth*. It predicted similar civilisational "overshoot and collapse," but this time on the basis of then-state-of-the-art computer modelling. Like Ehrlich's book, *Limits to Growth* was a runaway bestseller.

If the present growth trends in world population, industrialisation, pollution, food production, and resource depletion continue unchanged, the limits to growth on this planet will be reached sometime within the next one hundred years. The most probable result will be a rather sudden and uncontrollable decline in both population and industrial capacity.

There have been decades of criticism from numerous quarters quarrelling with the Malthusian assumptions of the modellers. Angus and Butler zero in on those of economist Christopher Freeman, the founder of the University of Sussex's Science Policy Research Unit, made just months after the report came out:

> What is on the computer print-out depends on the assumptions which are made about real-world relationships, and these assumptions in turn are heavily influenced by those contemporary social theories and values to which the computer modellers are exposed.

Angus and Butler ably extend Freeman's argument:

> What the model actually tries to do is to use neoclassical economic theory to predict how much economic growth will result from various levels of population growth, and then to estimate the emissions growth that would result ... In short, if your computer model assumes that population growth causes emissions growth, then it will tell you that fewer people produce fewer emissions.

Nevertheless, for all their repeated disconfirmation, and however much their logic leads inevitably and quite regularly to rather nasty, xenophobic lifeboat politics, the "Malthus in, Malthus out" conclusions of such reports, as Freeman describes them, have become a shibboleth of the green left.

3

To Infinity and Beyond!
(Or: The Myth of Carrying Capacity)

And what is the core erroneous assumption made not just by the *Limits to Growth* modellers, but all partisans of the politics of limits? Applying the concept of *carrying capacity* to humanity.

This population-biology term describes the speculative maximum, equilibrium number of organisms of a particular species that can be supported indefinitely in a given environment. A hundred-acre wood can support the consumption of a maximum of, say, 12 Eeyores. At this point, equilibrium is reached. Any more Eeyores would require more acres of wood, or each of the Eeyores consuming less.

Throughout the *Limits to Growth* report, the authors refer to humans having overshot our carrying capacity. From ex-World Bank economist Herman Daly to Al Gore, the doomsayers don't write "The end is nigh!" on their sandwich boards, but "We've overshot our carrying capacity!" Klein, Kunstler, Kingsnorth and the full coterie of anti-growth partisans repeatedly call attention to carrying capacity "overshoot." It's a reference, consciously or, more likely these days because the concept of carry capacity is so ubiquitous, unconsciously, to the influential 1973 book *Overshoot* by US environmental sociologist William Catton[16] that popularised the subject. The book is also Deep-Green-Resistance guru Derrick Jensen's bible.

"The carrying capacity of the planet," Jensen told an interviewer, urging people to read Catton, "is declining every day, because of the murder of the planet. Where I live, 180 years ago the rivers were literally full of salmon. Now the runs are much smaller. The entire world is diminished. The total weight of the fish in the ocean has decreased by 90 percent in the last 150 years or so. Obviously the planet can permanently support fewer humans now than it could before."[17]

Or, in Catton's words:

[O]ur lifestyles, mores, institutions, patterns of interaction, values, and expectations are shaped by a cultural heritage that was formed in a time when carrying capacity exceeded the human load. A cultural heritage can outlast the conditions that produced it. That carrying capacity surplus is gone now, eroded both by population increase and immense techno-logical enlargement of per capita resource appetites and environmental impacts. Human life is now being lived in an era of deepening carrying capacity deficit. All of the familiar aspects of human societal life are under compelling pressure to change in this new era when the load increasingly exceeds the carrying capacities of many local regions—and of a finite planet. Social disorganization, friction, demoralization, and conflict will escalate.

The attraction of the concept of carrying capacity is its simplicity. The number of variables is essentially two: the number of organisms and the amount of resources. For some reason, it's not Eeyores, but bacteria in petri dishes that tend to figure promi-nently as the metaphor given by anti-growth activists to help explain the carrying capacity concept further: there is a limited amount of medium (food) in the petri dish; bacteria consume the medium at a fixed rate; eventually the number of bacteria exceed the amount of medium available per capita and begin to die off.

(The UK's green left New Economics Foundation for its part prefers the metaphor of the 'impossible hamster'. The think-tank [which, I should say, even if it embraces degrowth economics, does a very great deal of good work otherwise] has produced a cartoon of a hamster that gets fatter and fatter until it is a 9-billion-tonne burping, growling Godzilla-rodent, careening through cities eating everything in its path.[18])

Problem number one with this model is that the petri dish is a classless society! There are no bourgeois and proletarian bacteria consuming resources at radically different and dynamically changing rates. It makes sense to talk of average consumption rates of bacteria, but no sense at all to talk of the consumption rate of "the typical Westerner," let alone "the average human." If you put me and Warren Buffett in a room together, the average net worth of people in that room is $29.1 billion. The 85 richest people in the world have as much wealth as the bottom half of all of humanity, according to development charity Oxfam. Only humans experience a *class society*.[19]

The richest seven percent of people are responsible for 50 percent of greenhouse gas emissions as of 2010. Meanwhile, as the aforementioned Angus and Butler show, the inflation-adjusted discretionary spending of a two-income family in the United States in the early 2000s was *less* than a single-income family in the early 1970s. And the top 20 percent of income earners account for roughly 70 percent of consumption. Angus and Butler present a trove of examples showing how the ultra-high-net-worth individuals consume fossil fuels and other resources at rates incomparable to the rest of us, north and south. People like Russian oligarch Roman Abramovich with his Bond-villain-esque 533-foot yacht with its own missile defence system and an escape submarine (no really); Indian petrochemicals titan Mukesh Ambani, with his 22-story mansion in Mumbai with nine elevators, 600 servants and an artificial snowfall room to escape the heat; and British retail magnate Philip Green, who "flew 300 guests to the south of France and put them up in a $1,600-a-night hotel so they could attend his son's bar mitzvah. He also flew in stone masons and other craftsmen to build a 300-seat synagogue on the hotel's grounds. The event featured an evening concert by Italian tenor Andrea Bocelli and another by pop star Beyonce and Destiny's Child."

So phrases such as "the greenhouse gas emissions of the average American" or "per capita consumption" contain absolutely no useful information.

In perhaps the sharpest insight in the book, taking on the anti-consumerist finger-wagging of Greenpeace and co., Angus and Butler write "We can reply to those who blamed BP's Gulf of Mexico disaster on our supposed addiction to oil: even if we accept the far-fetched idea that oil companies drill new wells only to please consumers, no one can reasonably suggest that consumers somehow forced BP to cut every possible corner, suborn regulators, or violate safety guidelines. Those decisions were made in BP's executive offices, and consumers had no say."

Most environmental crimes are committed on behalf of corporations, they note, but not by accident or *atypical* lawbreaking. Rather, "polluting is business as usual."

Unlike other organisms, who have instinct but no society, we can *alter* our social relations, indeed our whole political economy, completely transforming how we use the planet's resources. Our rate of consumption is malleable, dependent in part on something that is absolutely changeable: our *relations of production.*

To treat humans as no different from hamsters or bacteria is to forget all of social science, all of history, politics and economics. To forget what makes humans different from the rest of nature.

But we can also alter our consumption patterns radically in a second way that no other species can: we can invent new technologies. We are constantly transforming the *forces of production*[20] as well.

If through technological change, we can use half an ingot of steel to produce a widget instead of a whole ingot of steel, we can alter our rate of consumption of resources. Other organisms can only alter their consumption of resources by *speciating*—that is, becoming another organism.

For example, in *This Changes Everything*, Klein endorses the

Limits to Growth report, but she acknowledges that it got many things wrong and that its projections have not held up over time because the authors underestimated the capacity of "innovative technologies to unlock new reserves of finite resources." Nevertheless, she writes, "*Limits* was right about the most important limit of all" —natural sinks for pollution such as greenhouse gases. In other words, *Limits to Growth* was wrong because the authors neglected to include in their computer model humans' capacity to solve problems, but we won't be able to solve the problem of greenhouse gases.

Yet she never says what is so special about this new, apparently unfathomably inscrutable problem that forbids humans from solving it. As it happens, we already have a series of technologies that we *know* will solve the climate crisis, which I'll look at shortly. It is simply a lack of political will that is preventing their deployment.

Mine is hardly a new position though on the left. This is the very argument that Friedrich Engels mounted against Malthus in his *Outlines of a Critique of Political Economy* in 1844: "[T]here still remains a third element which, admittedly, never means anything to the economist—science—whose progress is as unlimited and at least as rapid as that of population."[21]

*

A brief comment on terminology here: Progressive critics of the market economy since Marx (whether they consider themselves to be Marxist or not) have often found it useful to employ his distinction between the *relations of production*—the sum total of socio-economic relations in a society—and the *forces of production*—the 'stuff' with which we produce things, i.e. technology, infrastructure, land, natural resources and our labour power. The *relations* and *forces* of production combine together in any epoch as the *mode of production*—the way that a society is organized as a totality, such as capitalism, feudalism, slavery-based economies, hunter-gather society, and so on. The reason these sort of concepts taken from heterodox economics, along with, yes, some concepts from classical liberal economics, are useful is that they allow us to understand the mechanisms and contradictions of an economic system, to understand *how* things happen, and as a result, most importantly, how we can change them.

I'm sorry for throwing some jargon at you, but by coming to grips with such concepts, we can begin to see how in altering our relations and forces of production, we can continue to grow *and* solve our environmental problems at the same time. Furthermore, these concepts show how by focussing on growth or scale of production, instead of the relations and forces of production, we may simply be recreating the problems that got us into this mess in the first place.

*

The mantra we keep hearing from the anti-growth advocates, "You cannot have infinite growth on a finite planet," seems so obviously true. Which is why it is so seductive to green activists. It's designed perfectly for a banner or placard. But it is only true if the rate of consumption is fixed, and we have just shown how in two clear ways — technological innovation and reorganization of our political economy — we can alter the rate of consumption. Indeed, the rate is constantly changing.

A more developed understanding of the finite and infinite is helpful for this discussion. Think of a single rubber ball. Like the Earth, it is bounded in the sense that very clearly there is an edge to the ball and there is only so much of it. It doesn't go on forever. It is not bound*less*. And there is only one of them. But it is infinitely divisible in the sense that you can cut it in half, then cut that half in half again, then cut that quarter in half, then that eighth in half, and so on. In principle, with this imaginary ball, you can keep cutting it up for as long as you like, infinitely extracting from this finite object.[22]

So long as we can keep innovating, changing the rate at which we use a resource, *in principle* and in our mathematical imagination any bounded lump of anything can be divided infinitely, even while being finite.[23] Thus, counter-intuitively, you *can* actually have infinite growth on a finite planet.

The kernel of truth to the anti-growth argument about finitudes here is that in the real world outside the perfectly, endlessly divisible rubber ball of our mind, things become much messier. Under capitalism, we don't sit around pondering Zeno's Paradoxes and complete versus incomplete infinities, but instead often inadvertently use much more of a resource regardless of whether we have a new technology that allows us to use less of it to produce the same number of widgets.

Why is this? What is the mechanism that drives this overproduction, this carry-on-regardless approach to resource use? The degrowthists will tell you that it is capitalism's requirement for growth. But this is far too crude a description of what occurs. What is fundamental is not growth per se, but capital's need for self-valorisation, or put another way, for self-expansion. A firm cannot receive back in value precisely what it has put in, otherwise it would go bankrupt. It must receive more value than it put in. Like a bicycle wheel, capital must keep on keepin' on or it will fall over. Superficially, this appears to be the same thing as growth, but it is not. And the distinction is not a philosophical trick, but a difference of profound significance. Here's why.

If the production of a commodity by a particular firm involves pollution of some description (and I'm using the term 'pollution' for simplicity's sake, but you can replace this term with 'extraction' or 'pumping out CO_2' or 'catching too many fish' or whatever it is that we have agreed is detrimental), and a rival firm produces a commodity that is identical in every way apart from the fact that it pollutes less and costs more, then the first firm will be more successful than the second every time. And if the second firm is consistently undercut by the first firm will not have self-valorised (expanded). The firm not be able to pay for new materials or labour or the upkeep of its machines and will go out of business. This is why capitalists, left to their own devices, have no choice but to pollute or extract or pump out CO_2 or catch fish at a rate that is heedless of what remains of our store of resources. It is not that they are evil or greedy. If one capitalist says to herself "To hell with the profits! The planet is more important!" then she will quickly be beaten by a rival who is not so scrupulous. To keep going, they will have to give up on such high-minded thoughts. And this is true regardless of size, whether a globe-rogering, \$11-bajillion-market-cap, Taibbian vampire-squid investment bank or a mom-and-pop corner shop that sells nothing but thimbles of rosewater-scented whimsy and hand-sewn felt puppets of characters from Wes Anderson films. If right next door, a big-box chain-store Whimsy-Mart opens up with vats of all-you-can-eat cut-price Owen Wilson dolls and that small business doesn't toughen up, then they're fucked.

There is a caveat to this though. If *all* firms that produce this particular commodity are regulated in such a way that only the less polluting (or non-polluting) production process is permitted by law, then there is no particular advantage or disadvantage for any one firm. But this is an intervention in the free market. It is a regulated capitalism. One could describe this perhaps as the liberal or social democratic response to pollution.

Another caveat to this story, this time in capitalism's favour, is that if a firm can produce a commodity that is less polluting using a production process that is *cheaper* than that of a rival firm, then of course the former firm will be the successful one. And this phenomenon—the pressure to produce a commodity more cheaply driving a firm to use fewer resources per unit of production—very much does happen.

The more critical, or socialist, opinion argues that however much we want to regulate capitalism, there will always be some new commodity or market inadvertently 'polluting' that has yet to be regulated. So the regulator is always playing catch-up. Further, capital's need for self-valorisation tends to strain at the leash of regulatory restraint, as there is always some jurisdiction where this regulation does not exist. Which means that there is a force in the economy constantly pushing toward pollution that we are forever trying to push back against, a beast we cannot tame or cage. This is why social democracy goes further toward preventing pollution than less regulated forms of capitalism, but cannot absolutely prevent the problem.

Similarly, while the trend of diminishing resource use per unit of production over time—a phenomenon known as 'decoupling'—is very real (and more on this shortly), it is not true at all times for all commodities or always at sustainable rates, thus decisions inadvertently leading to polluting overproduction continue to happen.

A democratically planned economy, however, would make production decisions on the basis of use-value—that is, on their utility to society—rather than just letting capitalists chase capital self-valorisation willy-nilly. We can continue to grow, but in a rational, planned fashion, avoiding the problem of inadvertent overproduction. We can slow down or hold tight or rearrange production until new efficiencies from technological innovation (i.e. a change in the *forces of production*) are forthcoming. If a certain form of pollution offers less utility to society than not polluting in that way, then we simply do not do so. It is not that the beast is caged, as with social democracy—or at least as social democracy claims to attempt. There is no longer a beast to cage.

Thus the problem with capitalism is not economic growth, but *lack of planning*, and so our target should be the mode of production (capitalism), not growth itself. Our goal is a *democratically planned economy*, or at the very least a much more democratic supervision of private decisions on investment and production, not degrowth or a steady-state economy.

To put it another way:

- The capitalist says: There may or may not be resource limits, but don't worry about them! Innovation will come along in time! *Full steam ahead!*
- The green lefty says: Innovation can't save us! There's an upper limit to what humans can have / an upper limit on the number of humans. *Slam on the breaks!*
- The socialist says: Through rational, democratic planning, let's make sure that the innovation arrives so that we can move forward without inadvertently overproducing. And move forward we must, in order to continue to expand human flourishing. So long as we do that, there are in principle no limits. *Let's take over the machine, not turn it off!*

As I've stressed, carrying capacity can in no meaningful way be employed to describe humans because we are so qualitatively different in two fundamental ways from every other species on the planet: we can change our political economy and we can develop new technologies. But beyond these two differences with other organisms we also have to recognize that as a result of socio-economic dynamics, consumption patterns alter in other, sometimes unexpected ways, as we saw earlier with fertility rates declined inversely to economic development. Thus in the end, there are so many, many more variables to consider than just the number of organisms and the amount of resources.

So no matter how science-y sounding, arguments that we have extended ourselves beyond the Earth's carrying capacity are in fact out of step with science and the historic and prehistoric evidence. Thus, as a result of all these other factors that we've now explored, when someone asks what are the limits to growth, what is the maximum carrying capacity of the Earth with respect to humans, how much can we consume, we have to answer: Well, it depends. It depends on the current state of technology, on the state of the environment itself, on our political economy, on the distribution of wealth, on what we value, on our aesthetics, etc. It depends ultimately on what we want.

And depending on these other variables, we can come up with a very wide range indeed of different estimates as to the maximum human population of the planet (which is, as we have established, in effect identical to the maximum human consumption the planet can provide). Siegfried Franck of the Potsdam Institute for Climate Impact Research and his colleagues suggested in 2011 that the planet can sustain up to 282 billion people, but by using *all* the land. A save-the-forests-scenario can sustain 150 billion people. A scenario in which only pasture is cultivated to feed animals produces a figure of 96 billion people.[24]

But even these estimates are far from the most outlandish scenarios. As a purely academic exercise, Dutch agronomist Cornelis de Wit in 1967 came up with a figure of 1.67×10^{21} maximum people, (subsequently readjusted down to a measly 10^{20} folks[25]), assuming the limiting constraint to be the amount of carbon contained in the Earth being embodied in people containing around 12 kg of carbon per person. To be fair, these hundred quintillion people would all have to be cannibals. But all the maximum carrying capacity figures offered up under these different, extreme options—the save-the-forests scenario, the don't-save-the-forests scenario, the, ahem, *Soylent Green* scenario—depend on unquantifiable *value* judgments. You and I might not like the idea of a Cormac McCarthy-esque[26] world of a hundred quintillion cannibals, but the back-of-the-envelope calculations do at least appear to tell us that Earth has the capacity to carry such a load.

Tell me what sort of world you want to live in, and I can adjust the carrying capacity numbers for you accordingly.

*

The Actually Pretty Soft Hard-Limits of Planetary Boundaries

But what about these planetary boundaries we keep hearing about? "Your logical kung fu may be strong," I hear you say, "but the facts on the ground show that on almost every front, humanity is pushing multiple natural systems into crisis."

This is not false. Moving beyond the highly problematic *Limits to Growth* methodology, in 2009, a group of some two dozen earth system scientists proposed a new framework of hard ecological limits, this time termed "planetary boundaries," in which the researchers attempted to comprehensively track every aspect of humanity's transformation of the environment, from climate change to biodiversity loss, in an evidence-based manner. They wanted to establish an easily understandable guide to what was left of an optimum "safe operating space for humanity." Beyond these boundaries, the researchers said, there is a risk of "irreversible and abrupt environmental change" that would make the planet significantly less habitable for humans.

The researchers, led by Johan Rockström of the Stockholm Resilience Centre and Will Steffen of the Australian National University, and including former NASA climate scientist James Hansen and Paul Crutzen—the atmospheric chemist and originator of the term 'Anthropocene'—identified nine such boundaries. Beyond climate change, these are ocean acidification, ozone depletion, the nitrogen and phosphorus cycle (the 'biogeochemical boundary'), freshwater use, land use, atmospheric aerosols and the 'toxics' boundary—meaning persistent organic pollutants, heavy metals, plastics and radionuclides.

"Transgressing one or more planetary boundaries may be deleterious or even catastrophic due to the risk of crossing thresholds that will trigger non-linear, abrupt environmental change within continental- to planetary-scale systems," the researchers warned.

Due to the complexity of the Earth system, precise thresholds or tipping points are fiendishly difficult to determine, so Rockstrom and his colleagues instead speak of ranges, the bottom end of which is a particular boundary. And currently, two and a half of the boundaries—climate change, biodiversity loss and the nitrogen cycle (but not the phosphorus cycle)—appear to have been crossed.

The concept in other words tells us how much more radical alteration of vital ecosystem services we can put up with before we begin to threaten the survival of our species. It is a map of the Anthropocene, the aforementioned proposed new geological epoch where humans are the main agent of change on the planet.

This new framework, substantially more rigorous and grounded in evidence than the *Limits to Growth* report of the 1970s, has proven exceptionally popular, endorsed by a range of international organisations such as the European Commission and the UN Environment Programme, embraced from TED to the World Economic Forum. The UK science journal *Nature* published a special edition of the publication dedicated to the schema. A blossoming of symposia, conferences and additional reports building on the initial concept and fine-tuning the boundaries. From economist Jeffrey Sachs to noted architecture critic and artisanal cheese producer Prince Charles, it has been embraced by a panoply of global 'thought leaders' and most green NGOs. A 2011 best-selling book by British environmental writer Mark Lynas, *The God Species*, riffed on and popularised the concept. No international environmental conference, whether focussing on government representatives, business leaders or activist groups, will fail to feature mention of the concept.

But this is not to say that everyone stands in agreement on its meaning, its utility or how it is used. And there are many who question whether it is a worthwhile document at all.

Fundamentally, the debate centres on whether these boundaries represent rigid, permanent upper limits to human societal growth, or whether they are dynamic and the whole schema more a rule of thumb.

Even when it was launched, a number of researchers asked by *Nature* for commentaries reacting to the concept said that while overall it was "worthwhile and useful," they were also cautious about how to choose upper limits on environmental degradation and whether researchers know enough to choose the right parameters. To take the toxics threshold for example, few scientists are unconcerned about the effects of the wide variety of pollutants on biological organisms, yet at the same time, it seems incredibly difficult to quantify a *boundary* for this. The same can be said of the atmospheric aerosols.

Tipping points are rarely predictable, and some worry that the idea allows us, like an undergraduate with a month before a paper is due, to say "Oh, well, we've got lots of time before things go pear-shaped! We don't need to worry just yet."

An editorial in Nature accompanying the special issue on the topic warned readers: "The exercise requires many qualifications. For the most part, the exact values chosen as boundaries by Rockström and his colleagues are arbitrary. So too, in some cases, are the indicators of change." The journal worried that there was so far little scientific evidence to suggest that stabilizing long-term concentrations of carbon dioxide at 350 parts per million, the particular boundary Rockstrom and his colleagues had picked on the climate front, is the right target for avoiding dangerous interference with the climate system. One might add that this may be reasonable a target for our society to aim toward, but then that is a political decision rather than an objective scientific datum. "Nor is there a consensus on the need to cap species extinctions at ten times the background rate, as is being advised," the journal continued. "Such numerical values ... should not be seen as targets. If the history of environmental negotiations has taught us anything, it is that targets are there to be broken."

Indeed, one of the boundaries mentioned in the report, depletion of the stratospheric ozone layer, which protects biological systems from ultraviolet radiation from the Sun, is a problem that has been solved. We came dangerously close to crossing this boundary, but as a result of the Montreal Protocol, which banned chlorofluorocarbon (CFC) containing aerosol sprays, and their replacement by hydrochloroflurocarbons HCFCs, which are less damaging to the ozone layer, and ultimately HFCs, which do not damage it at all, we have turned back. The ozone layer is recovering. Through innovation we continued to advance while solving the problem we had caused.

Thus many who find the report useful overall argue that these 'boundary' points should not be written in stone, but viewed as a heuristic. "Ever since early humans discovered fire and the benefits of collaborative systems such as collective hunting and social learning, human systems, not the classic biophysical limits that still constrain other species, have set the wider envelope for human population growth and prosperity," argues ecologist Erle Ellis, who absolutely accepts the reality of climate change and other anthropogenic despoliation of ecosystems, but criticizes a Malthusian reading of Rockstrom's report. "It was not planetary boundaries, but human system boundaries that constrained human development in the Holocene, the geological epoch that we have just left. We should expect no less in the Anthropocene."[27]

He, and co-thinkers of his at the Breakthrough Institute, a US self-styled 'ecomodernist' think-tank that has run into accusations of conservatism[28] for its embrace of nuclear power, large-scale energy infrastructure, and genetic modification, argue that it is precisely through economic growth that humanity will be able to afford and develop the new technologies and infrastructure that will enable us to solve the problems highlighted by the planetary boundaries concept like climate change and biodiversity loss. (Despite the accusations of right-wingery however, the institute's leading figures have in reality long been involved in left-liberal causes including campaigning against Nike's use of sweatshop labour and advising the late left-nationalist Venezuelan President Hugo Chavez, and generally arguing that the bulk of the transformative infrastructure will have to be built through public-sector endeavour.)

Ellis argues that:

The 'planetary boundaries' hypothesis asserts that biophysical limits are the ultimate constraints on the human enterprise. Yet the evidence shows clearly that the human enterprise has continued to expand beyond natural limits for millennia. Indeed, the history of human civilization might be characterized as a history of transgressing natural limits and thriving.

For Ellis and his colleagues, the 'hypothesis' of planetary boundaries should rather be seen as a useful rough guide, a map, to inform policy-making rather than as a series of hard and fast limits, or worse, as ocean chemist Peter Brewer worried in a commentary in *Nature* on the concept: "just another stick to beat citizens with." Speaking of the nitrogen cycle in particular, Brewer wrote: "it is likely that a large fraction of people on Earth would not be alive today without the artificial production of fertilizer … food is not optional."

Moreover, the planetary boundaries report was presented upon publication to the general assembly of the Club of Rome and is widely seen as a 21st Century update to *Limits to Growth*. I don't want to condemn the authors via guilt by association with these earlier Malthusians. Nevertheless, this suggests that the authors do indeed intend the concept to be understood not as a rough guide to the state of the biosphere and where it's going, but, like its forebear, describing a set of immutable, upper bounds to human civilisation.

*

I want to be clear, I am very much not saying "Don't worry, be happy." I've written far too much about climate science over the years to not be profoundly worried about the very real problems that we face and the truly gargantuan task involved in overcoming them. The conservative attitude that declares human ingenuity will always come to the rescue in time is incorrect. We certainly need to work hard to solve these problems, and there are many vested interests who stand to lose a great deal as a result of some of the changes our civilisation needs to make, but these problems *are* solvable.

Both the Pollyannas and the Cassandras are wrong, and both stand in the way of social justice, the former by condemning us all to catastrophic climate change and the loss of other vital ecosystem services for the sake of profit; the latter by condemning us all to a hair-shirted existence and refusal of further human development due to a romantic, unscientific belief in a static, unchanging balance of nature.

Thus, with the above caveats in mind, the planetary boundaries concept is a worthwhile effort overall. It is always useful for the locomotive engineer to be warned that the train will crash if we run out of track. But let's lay down some more track.

4

The Great Primordial Flatulence of Doom

One of the most important popular concepts that we do have to ditch is the idea that there is some *natural balance* that mankind is intruding upon as a result of our growth. Ecosystems aren't static, only being pushed out of equilibrium by dastardly humans. The reality is that they are in constant flux, punctuated by brief periods of equilibrium. What is evolution, but a process of constant and sometimes sweeping change? A fascinating demonstration of this is the little-known story of the Great Oxygenation Event, one of the most monumental cataclysms life on Earth has ever known. Yet at the same time, if it weren't for this holocaust some 2.3 billion years ago, which has been described as the most catastrophic event in the history of cellular life[29], humans and most other complex life on the planet wouldn't exist.

Over the course of 200 million years in the young Earth's oceans, cyanobacteria—sometimes called blue-green algae—had steadily produced sufficient free oxygen to wipe out most of the planet's inhabitants, which at the time were anaerobic—organisms that don't need oxygen to grow. In the case of obligate anaerobic organisms, oxygen is toxic. Initially a small proportion of the young world's organisms, cyanobacteria were the first to have evolved a mechanism to split water molecules using energy from sunlight—photosynthesis—and churn out oxygen as *waste*. This impressively energy-boosting innovation gave cyanobacteria a tremendous leg up, but at the expense of almost everybody else. Obligate anaerobic organisms either went extinct or shuffled off to niches where oxygen was absent.

Even worse, all this additional oxygen 'pollution' reacted with methane in the atmosphere. Methane is a greenhouse gas some 72 times stronger than carbon dioxide, and as a result, had been keeping the Earth cosy and warm. The sharp drop in atmospheric methane concentration triggered a similarly sharp drop in average global temperatures and the first known, most severe and longest ice age, a snowball Earth episode that lasted some 300–400 million years. Also known as the Oxygen Catastrophe, Oxygen Crisis or Great Oxidation (I myself like to think of it as the Great Primordial Flatulence of Doom), this development was the first and one of the most thoroughgoing of the Earth's extinction events.

(Many legally-minded environmental groups are pushing for the adoption of 'ecocide' laws. In 2010, UK lawyer Polly Higgins, the chair of the Eradicating Ecocide Global Initiative, proposed to the United Nations in 2010 that ecocide be made an international crime defined by an amendment to the Rome Statute as "The extensive damage to, destruction of or loss of ecosystem(s) of a given territory, whether by human agency *or by other causes*, to such an extent that peaceful enjoyment by the inhabitants of that territory has been or will be severely diminished" [emphasis added]. One presumes that were the International Criminal Court around in Precambrian times, Higgins and her a panoply of juridical luminary backers would have the SS and Himmlers of blue-green algae put in the dock in The Hague.)

However, the disaster, which makes anthropogenic global warming appear as an amateur dramatic society production by comparison, also takes the name of the Oxygen *Revolution* because it accelerated evolution, forcing tremendous changes in metabolism. Most of the organisms that survived were now endowed with a radically increased free energy supply. This energy-harvesting bounty led to an explosion in more complex organisms, including eventually multicellular life. Fossil evidence suggests that the emergence of mitochondria, the engine room of the cell, evolved around this time. You can think of it like the X-Men, where characters are now bestowed with these amazing superpowers to do incredible things that no one before them could do. Yet make no mistake, this was Armageddon for everyone else. It's an X-Men film where Magneto wins. (Which would be awesome, by the way. Come on, who else thinks Professor Xavier is a bit of a pussy?) The extinction of almost all life at this point was the precondition of everything that came after. All life on Earth today can trace its family tree back to the handful of bacteria to survive the tribulations of the Great Oxygenation Event.

Likewise, the most recent mass extinction, the Cretaceous–Paleogene (K–Pg) extinction event (the artist formerly known as the KT Boundary Event), when a giant comet or asteroid some 10 km across hit the Earth, is famous for killing off Dino, Barney and the rest of the non-avian dinosaurs. But it also eradicated 75 percent of all plants and animals, opening up ecological niches for other species, and leading to an astounding diversification of mammals in particular. We would not be here if it were not for these and other previous ecological apocalypses.

When obligate anaerobic bacteria were displaced and aerobic life exploded, was this a bad or a good thing? It was sad for the anaerobes, I suppose, but members of the mitochondria fan club might have a different opinion. When most of the dinosaurs died off, was this a tragedy? I know I'd pay top dollar for a ticket to Jurassic Park if it were real, especially if they got Jeff Goldblum to record the audio-guide, but I am quite happy for humans to be around as a consequence of the K-Pg extinction event. The dinosaurs themselves were the beneficiaries of an earlier mass die-off, the Triassic-Jurassic extinction event, which allowed them to fill the ecological niches previously dominated by the crurotarsi. In total, geologists and biologists agree that there have been five 'great dyings' since life first emerged on Earth, as well as a series of lesser extinction events.

If for a moment we try to be more objective, or 'biocentric', and forget about what was most optimal for our species, but rather try to think about what is optimal for life itself—to abandon our 'anthropocentrism', as the deep ecologists would recommend—then can we genuinely say that aerobic life is superior to anaerobic life? Are horses better than t-rexes? Clearly, the words 'better' or 'superior' are meaningless here. All we can say is that the different organisms are, well, different. More complex, certainly, but that is something we as humans care about. The rest of life is indifferent to its level of complexity.

Today, many researchers argue that the planet is experiencing a *sixth* great dying, this time not caused by asteroids or volcanoes or supernovae or oxygen-farting bacteria, but by us. Researchers estimate that we have caused the extinction of around 1,000 species over the last 200 millennia that modern humans have been around, and another 20,000 are threatened with extinction according to the International Union for the Conservation of Nature. Other researchers argue that however dramatic the current die-off, losing a few percent of assessed species, this does not compare in magnitude to the Big Five extinctions—yet. It would take a few more centuries for this to happen.[30] Regardless, while our extinction event may not be the most extensive, it is certainly the fastest, clocking in at about a thousand times faster than the average extinction rate the planet experiences. The International Commission on Stratigraphy, the guardian of geochronological nomenclature, has set up a working group to assess whether our current era should officially be called the Anthropocene. And without some sort of significant intervention by us to reverse this trend, the likely sort of organisms that are to dominate will be, as science writer David Quammen argued in a famous 1998 essay in *Harper's* magazine, "Planet of Weeds," hardy, versatile organisms that are able to thrive in human-dominated environments: invasive and synurbic (urban-preferring) species such as raccoons, pigeons, rats, gray squirrels, coyotes, Africanized honeybees, geckos, houseflies, ragweed, cheatgrass, blackthorn and thistle.

Is there a larger pattern to these invasions? What do fire ants, zebra mussels, Asian gypsy moths, tamarisk trees, maleleuca trees, kudzu, Mediterranean fruit flies, boll weevils and water hyacinths have in common with crab-eating macaques or Nile perch? Answer: They're *weedy* species, in the sense that animals as well as plants can be weedy ... They reproduce quickly, disperse widely when given a chance, tolerate a fairly broad range of habitat conditions, take hold in strange places, succeed especially in disturbed ecosystems, and resist eradication once they're established. They are scrappers, generalists, opportunists. They tend to thrive in human-dominated terrain because in crucial ways they resemble *Homo sapiens:* aggressive, versatile, prolific, and ready to travel.

Paleobiologist Anthony Barnosky, for his part, calls this homogeneity the "McDonaldsization of nature."[31] But is this a good thing or a bad thing? Quammen presumes it to be bad; yet another example of humanity's evisceration of the natural world. But would a *Weed Revolution* be any different to an Oxygen Revolution? Is an abundance of coyotes less preferable than an abundance of the red foxes they may be displacing in many parts of North America? Red foxes themselves typically dominate and can displace other fox species such as the Arctic fox, and in Australia is one of the continent's most invasive species.

Since the 1980s, we have been constantly discovering organisms, often microbes, that exist in extreme environments previously thought to be thoroughly inhospitable to life: ridiculously hot, cold, dry, dark, acidic, alkaline, radioactive, pressurized, low in oxygen, high in heavy metals, intense gravity, or deep in the Earth, or some combination of these. Tardigrades for example, tiny invertebrate 'micro-animals' also known as water bears or moss piglets, are totally zen even up to 151 degrees Celsius or down to just one degree above absolute zero (at least for a few minutes). They can survive without water for at least 10 years and, as temperature descends, their body composition shifts from 85 percent water down to just three percent. A tardigrade can withstand up to 6000 atmospheres of pressure and is the first known animal to survive the vacuum of space. Likewise, doses of radiation a thousand times above what would be lethal for most other animals don't make no moss piglet squeal.

We are confronted with the intriguing question of whether life may even be an intrinsic property of chemical reactivity itself—a question asked by the US National Research Council's Committee on the Limits of Organic Life in Planetary Systems when it was tasked in 2007 with considering whether there could be life on Saturn's largest moon, Titan. That is to say that life could be an inevitable result of chemical reactions involving stable carbon-containing molecules. Until we have more than one data point to consider, by discovering life on some other world, we simply cannot say. What we can say is that whatever the true level of the fragility or robustness of *homo sapiens*, life itself is profoundly resilient. Environmentalists are not 'saving life on Earth', which will do very fine with or without us, thank you very much.

Paul Kingsnorth encourages us to ditch the anthropocentric viewpoint and embrace the ecocentric viewpoint in his 2013 essay "Dark Ecology," in which he describes five concrete actions that, along with building refuges from "the ongoing collapse of social and economic infrastructures," are not, according to him, wasting your time. Fourth out of the five is: "Insisting that nature has a value beyond utility. And telling everyone. Remember that you are one life-form among many and understand that everything has intrinsic value. If you want to call this 'ecocentrism' or 'deep ecology,' do it." However radical Kingsnorth's eco-pessimism may be, such calls to ecocentrism or biocentrism, a nature-centred instead of human-centred ethics (a nonsense to begin with, as humans are not separate from nature), are very common indeed, right into the mainstream.

Yet if life is so extremely robust (with us likely soon finding whether this is true even beyond the Earth), and if life itself cares not a jot which species live and which species die, then when we say that we are horrified by the idea of a Weed Revolution, then this is, counter-intuitively, in fact an *anthropocentric* assessment, not a biocentric or ecocentric one, which we are making solely on the basis of utility to humans. The biocentric viewpoint should not care about the weedification of life. The fossil record suggests that in the wake of *any* extinction event, weedy species proliferate, with life re-diversifying after about five to ten million years. So this has happened before, yet life keeps on keepin' on, blithely aloof about the level of biodiversity.

Concerns about biodiversity are in fact *anthropocentric* concerns. We should care when species go extinct not because of their intrinsic worth, as Kingsnorth suggests, but because the loss of species means a decline in the effectiveness of the services that living systems provide to humans, such as filtering of air and water, fixing of nitrogen, cycling of carbon, prevention of floods, and pollination of crops. (As we learn more about how the advanced cognitive capabilities of some animals, such as the Great Apes, cetaceans [whales and dolphins] and elephants, are similar to those of young humans, we may decide to begin to see members of these species as having intrinsic worth. We are already beginning to have a discussion about according such creatures additional protections. I am broadly sympathetic to such an argument, although suspicious of according *human* rights to animals that are not human. But this is a question of animal ethics, a topic beyond the remit of the current discussion and a topic for another time.) Biodiversity loss diminishes the resilience of these systems whose functions we depend upon. Most people think that the loss of keystone predators such as tigers or sharks is a tragedy because future generations of humans will not be able to experience their majesty. The maintenance of such 'charismatic megafauna' is certainly a worthy goal in itself. A world that is less beautiful to human eyes is a diminished one (although the corollary to this is that if we can make it more beautiful, then we should do so). But more concretely, keystone predators tend to assist with relative ecosystem stability. Without such creatures, humans need to be more active in expensive management activities to keep ecosystems in conditions optimum for human needs.

Put another way, a weedy Earth is just fine for the weeds, but it would be less useful to us. The displacement of the red fox by the eastern coyote in Massachusetts appears to have resulted in a spike in the number of small tick-bearing mammals such as mice, shrews and chipmunks, which in turn produces an increase in ticks that carry Lyme disease. Again, life itself is indifferent to this development; it is we humans that are not particularly happy about it.

Finally, for all that we have discovered about the rest of nature, most of it remains a vast library of unread, even unknown books, full of undiscovered medicines, materials, tools and models for engineering. Biodiversity loss is thus an act of cultural vandalism and loss of knowledge greater than the burning of the Library of Alexandria.

This does not mean that we should attempt to preserve all species forever. Extinction, like death, is a vital part of life. An attempt at a perfect preservation of all species would be a terrible misadventure, limiting the development of new benefits to humanity. Meanwhile, there are some organisms that we actively want extinguished. Leaving aside the active debate over whether viruses should be considered living organisms (the debate arises from facts such as that they cannot reproduce independently, are metabolically inert, and contain no cell structures, yet they do contain genes and evolve via natural selection), does the smallpox virus have intrinsic value? Should we not have eradicated this wretchedly disfiguring and murderous disease that took the lives of between 300 and 500 million people over the course of the 20th Century? Through scientific advance, we now live in an era where we are having a discussion as to whether we should eradicate mosquitoes, the greatest killer of humans.

As we saw with the cyanobacteria, even radical, catastrophic environmental alteration cannot be judged to be bad or good outside of its consequences for humans. As Mark Davis, a biologist and leading critic of the current war on invasive species, says, "There isn't such a thing as a healthy ecosystem or a sick ecosystem. Ecosystems are just out there. There's no particular goal or purpose. They're just the species and the physical and chemical processes taking place."[32]

The organisms that thrive in the ecosystem of an oil spill such as oil-consuming extremophile bacteria are not intrinsically less preferable to the organisms whose populations they have displaced. An oil spill is unwanted because it diminishes human uses of that ecosystem.

We must indeed in the end say that the Oxygen Revolution *was* good, but only because it increased the complexity of life, and this was only good because it further opened up the possibility of the emergence of humanity, the pinnacle of complexity in the known universe. Life itself does not assess increased complexity as morally better than simplicity.

Embracing a biocentric approach would actually entail saying "Bring on the Sixth Great Extinction!" because extinction events are a perfectly normal part of the grand tumult of the evolutionary process, happening quite regularly every 26–30 million years long before we arrived on the scene. If could even be cheekily argued that preventing mass extinctions is a human intercession in the 'natural order of things'. I'll not go there though, and simply say that we do not want a sixth, anthropocene great extinction and should actively intervene to prevent this, or anything else that would threaten such an event, not for life's sake, but for our own.

It might seem strange to say this as a nature lover, but from all this, it follows ineluctably that *it is only through appreciation of humans that there can be an appreciation of the rest of nature.* We care about nature because we care about ourselves and our desires. Prior to the advent of humans, nature was indifferent to the particular form that it took.

So to wish for an end to humanity so that the rest of nature can thrive—as do those deep greens who describe the human race as 'a cancer on the planet' or those in the Earth First! Journal writing under the pseudonym Miss Anthropy in the late 1980s and early 1990s that AIDS was an environmentalist's dream come, applauding nature's way of cleansing itself of overpopulating humans—is not merely vile, but *nonsensical.*[33] Life does not care whether we are here or not. Only we do.

Thus in any discussion of climate change, pollution and biodiversity loss, the goal cannot be maintaining pristine nature or some mythical natural ecosystem balance, but instead maintaining an ecosystem that is *optimal for humans.*

Or, to put it even more bluntly, the goal can only be to maximise human flourishing. Maintaining a certain average global temperature, i.e. the sort of temperature that permitted unprecedented human flourishing in the Holocene—the warm geological epoch that has existed since the end of the last ice age—is thus simply instrumental toward the central goal (ever greater human flourishing), and not an end in itself.

*

The concept of a 'balance of nature'—that, left to its own devices, nature returns to an equilibrium state—is as old as human history. Ancient Egyptian cosmology was centred on the idea of Ma'at, of order, justice, stability and a return to balance of all things out of balance. Personified in the winged goddess of the same name, Ma'at was the great regulator, making the sun rise, causing the seasons to turn, and at the moment of creation rescuing the universe from chaos and placing it in perfect equipoise. At the end of a person's life, his heart was placed on a scale, balanced by Ma'at herself, a judge in the underworld as well as a god.

It is understandable that we thought this way for so long: at the scale of a human life, the only things that one could notice changing significantly were the seasons, and they of course were cyclical, and so there was a balance there too. The grand, lurching, disruptive dynamism and constant change of evolution exists—at least on a noticeable level—on timescales much greater than that of a single human life, however long lived. It is only science that has opened up such scales so as to be viewable by us.

The idea dominated conservation work for decades, and we still hear green campaigners regularly declaring that nature knows best, that humans are parasites, and that were the wild world left to its own devices, all would be well. The Bolivian government of Evo Morales, which has done so much to improve the lives of ordinary Bolivians after 500 years of some of the most acute injustice in the Americas, has instituted a set of seven 'Mother Earth' rights into the country's constitution, including the 'right' to equilibrium and to restoration.

"There still exists the assumption that nature has appropriate states to which it should return, and these states are perceived as the 'natural' outcome even if their occurrence would contradict either theoretical predictions or existing data on predator–prey dynamics," wrote Corinne Zimmerman and Kim Cuddington in a 2007 paper in the *Public Understanding of Science* journal on the continued embrace of the balance-of-nature concept by under-graduate science students long after their professors had abandoned the idea. "For example, it is often assumed that predators 'control' herbivores, rather than eradicate species, regardless of situation. The assumption that such species extinc-tions never occur is contrary to both data and theoretical models."

Because the idea of a balance of nature is so deeply embedded in our culture, in our worldview, and chaos and change so terrifying, unpredictable and difficult to wrangle, it is understandable that so many of us would feel profoundly that it must be true.

Nevertheless, it is not. Long before the end of the last century really, the concept had been abandoned by ecological scientists as cumbersomely value-laden, romantic, religious or mystical even, and very much an inaccurate description of what is in fact eternal chaotic, dynamic turmoil.

Ecosystems can be and regularly are destabilized in the absence of human influence. Even the remaining wild parts of the world popularly thought to be pristine and stable, such as South American rainforests or the Serengeti savannah, are in fact in constant flux. Remember that the vast bulk of extinctions occurred before humans even evolved.

Mother Nature is not beneficent. She is not maleficent either of course. Merely augustly, majestically capricious.

As plant ecologist Steward Pickett of the Institute of Ecosystem Studies of the New York Botanical Garden told the *New York Times* 25 years ago, the balance-of-nature doctrine "makes nice poetry, but it's not such great science."

In fact, given the fact that we are now transforming the environment to such an extent that as a result of the actions we have already taken we may induce runaway global warming, if we do actually try to wait like Godot for natural balance to return when it will not, it is there that catastrophe awaits. Earlier case studies have shown that when simplistic balance-of-nature doctrine informs conservationist practice by for example policy-makers, park rangers or fisheries managers, the particular natural equilibrium outcome expected may not materialize. Theoretical and population ecologist Kim Cuddington of the University of Waterloo describes the unforeseen result of a decision to introduce wolves to Isle Royale National Park in Lake Superior:

> In the 1940's the moose on the island were outstripping their food supply, the park personnel feared a population crash was imminent. They reasoned that moose populations were too high because there were no natural predators on the island, and decided to introduce wolves in order to restore a balance. Clearly, these ecologists thought that the interaction between moose and predators would create a lower moose density. But equally clearly, the park managers did not imagine that wolves would drive the moose population to extinction, nor was it likely that they imagined a very large amplitude oscillation of wolves and moose would occur; after all, the moose were the tourist attraction they were trying to preserve. The equilibrium that the park managers wanted to create was of a very specific kind. Why did they imagine the wolf introduction would create such a balance?

Researchers have been investigating the relationship between wolf and moose populations on the island for more than fifty years now, a fascinating long-term study—indeed, it is the longest continuous investigation of a predator-prey system in the world.

"Important attitudes about how we should relate to Nature, and some of our abusive relationships with Nature, are rooted in convictions that we understand Nature well, and can accurately predict how Nature will respond to our actions," the researchers write, describing their project to predict and understand a relatively simple natural system.

> But the more we studied, the more we came to realize how poor our previous explanations had been. The accuracy of our predictions for Isle Royale wolf and moose populations is comparable to those for long-term weather and financial

markets. Every five-year period in the Isle Royale history has been different from every other five-year period – even after fifty years of close observation. The first 25 years of the chronology were fundamentally different from the second 25 years. And the next five decades will almost certainly be different from the first five decades. And the only way we will know how, is to continue observing. The most important events in the history of Isle Royale wolves and moose have been essentially unpredictable events – disease, tick outbreaks, severe winters, and immigrant wolves.

Rich, dynamic variation, not 'balance of nature', seems to be the force that guides nature. Natural history might be much like human history – explainable, but not predictable.

In a similar vein, in an essay critiquing agro-ecological farming— an alternative agriculture movement that attempts to design farming systems that 'mimic nature'—agronomist Andrew McGuire argues that if there is such a thing as nature's wisdom at all, it occurs only as the result of evolution's optimisation of species to their ecological niche, not the optimisation of ecosystems, and as a result, we should not be afraid of improving on nature.

If what we see in natural ecosystems is not optimized, but random (stochastic, say the ecologists), we should be able to do just as well or better. We can, with ingenuity, wisdom, and a good dose of humility, purposefully assemble systems that outperform natural ecosystems in providing both products and ecosystem services...

By taking advantage of individual species' properties and processes, and by managing abiotic conditions (soil physical and chemical properties and water levels, etc.) we can create designer agro-ecosystems, successful by criteria that matter in agriculture; productivity, efficiency, and stability. *I propose that this is, in fact, what we have been doing all along* … and that the "balance of nature" has only been a distraction from our efforts to improve the sustainability of our agriculture, a distraction that should be decisively cast aside [emphasis added].

Thus while there are certainly many good reasons to restrict the damage that our activities cause to ecological services, phrases such as "protecting the land" or "saving the planet" are ultimately empty of content. When we alter the environment, some of these changes have a negative impact on human welfare and some of them have a positive impact. The slogan on environmentalists' placards should be "Avoid self-harm!" not "Tread lightly on the Earth!"

We must push onward through the Anthropocene, fixing the problems of climate change, biodiversity loss and pollution not to return to some mythical balance, not for the rest of nature's sake, but for our own.

*

I'll go further still, and accuse such vulgar environmentalism of making precisely the mistake that activists accuse industrial civilisation of doing: drawing a false distinction between man and nature, expressed concisely by pioneer environmentalist and social anarchist Murray Bookchin in 1974: "I am trying through ecology to heal the wound that was opened by humanity's split with nature thousands of years ago."

In one respect, this distinction is obvious and absolutely correct. Humans are radically different from all other organisms. No other animal has our level of cognition, our self-awareness, our capacity for language, technology, art, abstract reasoning, or, of course, fire. We alone can choose our destiny. We alone make history.

But simultaneously, for all our uniqueness within nature, we remain *within* nature. We are not separate from it.

Our skyscrapers are not separate from nature; they *are* nature, as much as a termite colony's cathedral mound or a chaffinch nest or a bee hive. As are our iPhones and washing machines and bathyscaphe research submarines that take us to the depths of the Mariana Trench. Even the Voyager I space probe, 127.74 astronomical units from Earth at the time of writing, having crossed the heliopause and entered interstellar space on 12 September 2013, is nature, because it is of us, and we are nature too.

When Derrick Jensen writes: "If we wish to stop the atrocities, we need merely to step away from the isolation. There is a whole world waiting for us, ready to welcome us home," I am sympathetic to his lament for the self-harming destruction wrought upon the environment, but there is no home in nature to return to. We never left.

When environmentalists remonstrate society for "forgetting where we came from," for "losing touch with nature," the response must be: "Why, in fact it is *you* who has forgotten what nature is."

It is true that humans are the most dramatic of ecosystem engineers, far more capable of environmental transformation than beavers or worms or corals. But every bit of that dramatic transformation has been done by nature to herself. If cyanobacteria oxygen flatulence is natural, how less so are BP oil spills? Of course nature is now in the process of considering whether what she has done is very sensible, whether oil spills should be prevented by bring an end to the era of fossil fuels. Nature is publishing articles in academic climate-change journals and chaining herself to logging trucks and debating with herself whether to build nuclear power plants or offshore wind farms or both.

We are nature, and all that we do to nature is natural.

5

In Defence of Stuff

"At its core, it is a crisis born of overconsumption by the comparatively wealthy, which means the world's most manic consumers are going to have to consume less," writes Naomi Klein in *The Nation*.[34] Who are these 'comparatively wealthy' folks? Oh, everyone in the West. Once again, for someone who claims to tilt against capitalism, it is remarkable that Klein makes no class-based differentiation in developed countries. We all apparently consume at the same rate. And this has to come to an end. "We weren't born having to shop this much, and we have, in our recent past, been just as happy (in many cases happier) consuming far less."

It's the same anti-consumerist argument that was at the heart of her first book, *No Logo*, and it's also the ideology that animates the infamous Reverend Billy and the Church of Stop Shopping, with his church-revival-style protests outside the Disney Store and Walmart and warnings of the 'Shopocalypse' that have been featured in three different biographical documentaries.[35] Books about anti-consumerism are almost as popular as books about the apocalypse: There's *All Consuming* by Neal Lawson; *Affluenza* by Oliver James; *Affluenza: The All-consuming Epidemic* by John De Graaf and David Wann; the updated and expanded edition, *Affluenza: How Overconsumption is Killing Us and How to Fight Back*; and a suspiciously similarly titled book called *Affluenza: When Too Much is Never Enough*, but this time by Clive Hamilton, "the best-selling author of *Growth Fetish*." There is *How Much Is Enough?* by Allan Durning; *The Overspent American: Why We Want What We Don't Need* by Juliet Schor; *The Joy of Less: A Minimalist Living Guide* by Francine Jay; *The High Price of Materialism* by Tim Kasser; *The Simple Life: Plain Living and High Thinking in American*

Culture by David Shi; *The Paradox of Choice: Why More is Less* by Barry Schwartz; *Consumed: How Markets Corrupt Children, Infantilize Adults, and Swallow Citizens Whole* by Benjamin Barber; and *Enough: Breaking Free from the World of Excess* by John Naish. I could go on. There's *Adbusters* and *Real Simple* magazine and the peer-reviewed *Journal of Consumer Culture* that "takes a critical perspective on consumerism." In the wake of the wave of centre-left governments elected in the 2000s in much of Latin America, northern activists have worked to popularise the Andean concept of *Buen Vivir* in Spanish (Castilian), or *'sumak kawsay'* in Quechua and *'suma kamaña'* in Aymará,[36] which emphasises that what is important is to "live well, not better,"[37] and alights upon the incorporation of degrowth-like concepts into the Bolivian and Ecuadorean constitutions. The two countries are also celebrated for convening the People's World Conference on Climate Change and Mother Earth Rights in 2010, with Bolivia in the same year introducing the *Ley de Derechos de la Madre Tierra* or 'Law of the Rights of Mother Earth', a bill that gives Mother Earth, or 'Pachamama', a legal personality who can launch judicial action—through her temporal representatives, humans—in defence of her rights.

Anti-consumerism has become a fundamental doctrine of the modern left, indeed of mainstream thought across the board.

But the leading guru of the anti-consumerist philosophy has to be Annie Leonard, the author of the book and viral cartoon *The Story of Stuff*,[38] and appointed in 2014 as executive director of Greenpeace USA. Her 20-minute animated documentary was released in 2007 and has been watched by 30 million people since, translated into 15 languages and is regularly shown in elementary schools, churches and during corporate sustainability trainings. (In the process of researching this book, I discovered yet another book, *The Secret Life of Stuff* by Julie Hill, which is almost identical in content and even book design, and *Stuff: The Secret Lives of Everyday Things* by John Ryan and Alan

Thein Durning of Northwest Environment Watch. The anti-stuffists may think that there's too much stuff, but they clearly aren't very worried that there's too much stuff being written about stuff.)

You'll be familiar by now with many of the cartoon's arguments, presented by Leonard, on how if everyone consumed at US rates, we'd need three to five planets; how "the average" house size has doubled since the 1970s (ignoring again in both cases the variation in consumption by different classes within the US. Is she even familiar with the housing crisis in cities like London, New York, Paris and Vancouver, where due to property speculation and an end to public-housing building programmes, millions of young people and families are spending more than half of their income trying to squeeze into tiny, crumbling, mildew-ridden apartments where nothing works?[39]); and how "you cannot run a linear system on a finite planet indefinitely." But we've already dispensed with these sort of positions. So I'd actually like to make a robust, unapologetic defence of 'stuff', of consumerism, from the left. I want to show how anti-consumerism should be immediately recognized as an alien and antagonistic ideology to socialists, social democrats and liberals.

*

Early on in the cartoon, Leonard recounts a genuinely disturbing statistic: "30 percent of the kids in part of the Congo have dropped out of school to mine coltan, a metal we need for our cheap and disposable electronics."

That's terrible. It really is. But the reality is that coltan is not just used in iPhones and PlayStations, but in almost every kind of electronic product, including the computers that were used to make the *Story of Stuff* cartoon and upon which it spread virally via Facebook, Twitter and YouTube, and the computers that are used for everything from optimising energy efficiency in buildings to modelling climate change. Coltan also happens to be used in air and land-based turbines. Are turbines more 'stuff' that we don't need?

Are wheelchairs, dialysis machines and computers that are used to model the misfolded proteins of people with neurodegenerative disorders 'stuff' that we don't need? (I wonder sometimes how disabled-rights activists feel about their technophobic green left allies at times. When I was a student activist with the Canadian Federation of Students years ago, a group of anarchist-leaning delegates at a national conference got up and demanded that the organisation stop wasting money on professional catering. Instead, we should all go dumpster diving, they said. One wiseacre who had little time for this sort of more-radical-than-thou nonsense immediately responded: "But aren't you worried that this isn't very wheelchair-accessible?" And the anarchists, mortified at their own unconscious assumption of able-bodied privilege, quickly withdrew their proposal.) To be fair, rigid primitivists like Derek Jensen would callously say yes, get rid of it all, let billions die, while Leonard would probably make a distinction between 'stuff' we need and 'stuff' that we don't.

But what is wrong with the communications and leisure benefits we derive from iPhones and PlayStations anyway? Having a device in your pocket that serves up to you much of the corpus of human knowledge is a tremendous advance. Quite soon, I am very confident, it will be like a medical tricorder from Star Trek, able to instantly diagnose ailments. How is this a bad thing?

There will always be new devices after whose invention we will say: How did we ever live without this? But by definition, this is *more* stuff. There is no point in history where we can say: "Okay, we've got enough now. Let's stop." It follows then that the progressive argument must be to say yes, we do want more stuff, and to argue against those calling for less stuff.

*

Pro-Trinket, Pro-Bauble

But what exactly is wrong with gaming consoles, OhMiBod dildos that plug into an iPhone, or Hello Kitty 40th Anniversary plastic toys in Happy Meals anyway?

In the wake of the Black Friday sales after US Thanksgiving that in recent years have begun to take place in other countries as well, or Boxing Day sales the day after Christmas in Commonwealth countries, where people line up (or queue) before dawn in the freezing November weather outside the local MegaMart for ridiculously cut-price deals on everything, I've begun to notice a welter of Facebook status updates, tweets and 'news' articles sneering at videos of the trampling, stampeding chaos and images of people coming to blows over 40-inch plasma TVs, lap-tops or tumble dryers.

A survey of the incomes of those racing through the aisles to get to that hundred-dollar stereo that normally sells for $400 should give the smug tut-tutters pause though. This is one of the few times of the year that people can even hope to afford such 'luxuries', the Christmas presents their kids are asking for, or just an appliance that works. In a democratically controlled economy, we may collectively decide on different production priorities, but surely we would still organise the production of items that bring people joy. Why shouldn't people have these things that bring them pleasure? Is the pleasure derived from a box-fresh pair of Nike running shoes or a Sony PlayStation 4 inferior to the pleasure the subscribers of *Real Simple* magazine derive from their $2000 coffee table made from recycled traffic signs? Likewise, why is the £59 hand-carved walnut locomotive from a Stoke Newington toy shop any less consumerist than the free plastic Elsa doll from Disney's *Frozen* accompanying a Subway Fresh Fit Kids Meal?

The difference is a poor-hating snobbery and nothing more. We see this when fashion designer and Labour-supporter-turned-Tory-turned-anti-capitalist Dame Vivienne Westwood petitioned Downing Street calling for a ban on food from genetically modified organisms. Challenged by a BBC reporter that not everyone can afford to eat organic, she replied that they should "Eat less" and stop getting fat.[40] *Qu'ils mangent de kale.* We see this in multi-millionaire celebrity chef Jamie Oliver's obsession with stopping poor people from eating 'unhealthy' ready meals, fast food and, here it is again, from getting fat. Too many poor families, according to Jamie, are just "eating chips and cheese out of Styrofoam containers, and behind them is a massive fucking TV."[41] His pet hate is the big supermarkets, preferring people somehow switch from a weekly shop at Tesco's to fitting into their schedules of work and family a daily trip to the farmer's market or independent grocer's: "From a supermarket you're going to buy a 200g bag of this or a 400g pack of that. If you're going past a market, you can just grab 10 mange tout for dinner that night, and you don't waste anything."

Publicizing her 2014 book, Naomi Klein told that bastion of anti-consumerism, *Vogue* magazine: "We all must learn to stop buying so much, which means learning to stop defining ourselves by what we buy."[42] For Klein though, there are of course some exceptions when foreswearing consumption. From the same article, the world's leading women's fashion magazine:

Although she may be the world's most famous critic of consumerism, she understands the joys of shopping. At an appearance in London, somebody asked her to name one thing she liked about capitalism. She instantly replied, 'The shoes.' ... In a neutral cotton shirt and tailored black pants she greets me on a front porch lined with sneakers in Papa, Mama, and Baby sizes, and leads me past shelves overflowing with children's books, past the kitchen with the Breville juicer that Klein uses with amused devotion. 'When I feel my blood sugar getting off,' she says, with a wicked smile that reminds me of the actress Catherine Keener, 'I drink a glass of kale juice. It's so disgusting you don't want to eat anything!'

"There will be things we will lose, luxuries some of us will have to give up," Klein adds in *This Changes Everything*. But not, apparently, the 400-dollar juicers.

The elitist hypocrisy inherent in this discourse was first highlighted by cultural critic Thomas Frank over a decade ago, who argued in a series of seminal essays in the 1990s curmudgeonly lefty journal *The Baffler* that the hip, pseudotransgressiveness of anti-consumerism and counterculture, ironically, is a major *driver* of consumption: "Business is amassing great sums by charging admission to the ritual simulation of its own lynching." This essential argument was further developed by Joseph Heath and Andrew Potter in their 2004 polemic against Kleinian, Adbusterian 'culture-jamming', *The Rebel Sell*:

Many people who are, in their own minds, opposed to consumerism nevertheless actively participate in the sort of behaviour that drives it. Consider Naomi Klein. She starts out *No Logo* by decrying the recent conversion of factory buildings in her Toronto neighbourhood into "loft living" condominiums. She makes it absolutely clear to the reader that her place is the real deal, a genuine factory loft, steeped in working-class authenticity, yet throbbing with urban street culture and a "rock-video aesthetic." Now of course anyone who has a feel for how social class in this country works knows that, at the time Klein was writing, a genuine factory loft in the King-Spadina area was possibly the single most exclusive and desirable piece of real estate in Canada.

The anti-consumers—an army of tattooed-and-bearded, twelve-dollar-farmers'-market-marmalade-smearing, kale-bothering, latter-day Lady Bracknells—just don't like the sort of people who like McNuggets. Westwood, Oliver, Klein and the rest of the finger-wagging brigade simply do not think these other, lesser consumers are making the right, sensible choices. Anti-consumption politics almost always seem to be about somebody else's wrong, less spiritually rewarding purchases. It is perhaps the pinnacle of conspicuous consumption. At the very least, no one should mistake this lip-pursed *bien-pensant* middle-class scolding for speaking truth to power.

And lamentably, as this passage from 1937's *The Road to Wigan Pier* by the great socialist journalist George Orwell reminds, such elitist tut-tutting from the liberal left has long been with us:

The miner's family spend only ten pence a week on green vegetables and tenpence halfpenny on milk (remember that one of them is a child less than three years old), and nothing on fruit; but they spend one and nine on sugar (about eight pounds of sugar, that is) and a shilling on tea. The half-crown spent on meat might represent a small joint and the materials for a stew; probably as often as not it would represent four or five tins of bully beef. The basis of their diet, therefore, is white bread and margarine, corned beef, sugared tea, and potatoes–an appalling diet. Would it not be better if they spent more money on wholesome things like oranges and wholemeal bread or if they even, like the writer of the letter to the *New Statesman*, saved on fuel and ate their carrots raw? Yes, it would, but the point is that no ordinary human being is ever going to do such a thing. The ordinary human being would sooner starve than live on brown bread and raw carrots. And the peculiar evil is this, that the less money you have, the less inclined you feel to spend it on wholesome food. A millionaire may enjoy breakfasting off orange juice and Ryvita biscuits; an unemployed man doesn't. Here the tendency of which I spoke at the end of the last chapter comes into play. When you are unemployed, which is to say when you are underfed, harassed, bored, and miserable, you don't want to eat dull wholesome food. You want something a little bit 'tasty'. There is always some cheaply pleasant thing to tempt you. Let's have three pennorth of chips! Run out and buy us a twopenny ice-cream! Put the kettle on and we'll all have a nice cup of tea! That is how your mind works when you are at the P.A.C. level. White bread-and-marg and sugared tea don't nourish you to any extent, but they are nicer (at least most people think so) than brown bread-and-dripping and cold water. Unemployment is an endless misery that has got to be constantly palliated, and especially with tea, the Englishman's opium. A cup of tea or even an aspirin is much better as a temporary stimulant than a crust of brown bread.

Meanwhile, for all *The Story of Stuff*'s apparent anti-corporate rhetoric, its solution to those poor Congolese kids is not stronger labour rights, but in effect a boycott, a retreat from purchasing these gadgets. We can see now how such a film can adapt itself quite comfortably to corporate sustainability workshops in a way that a documentary on union organizing never could. No Sweat, the UK-based anti-sweatshop campaign group, explicitly argues against a boycott of companies like Apple, Nike or Disney that outsource production to Chinese or Bangladeshi sweatshops, arguing that as appalling as such jobs are, having them is better than what these kids would experience without them. The best solution is not Buy Nothing Days, but to campaign for labour rights domestically and, if you're not in a union, to join one. Workers in the developing world benefit the more workers everywhere are unionized.

An anti-consumerist model of campaigning simply and ineffectively replaces that of a trade unionist model.

*

"It wasn't always like this," Leonard adds. "Ask your grandma. In her day, stewardship and resourcefulness and thrift were valued."

Recently, I was at a local lefty public meeting on inequality where one audience member said: "Everything was basically fine when my grandparents were young. We knew who grew our food and how to fix things and we didn't have iPods. We don't need all this stuff."

"The primary way that our value is measured and demonstrated is by how much we consume ... and do we! We shop and shop and shop," Leonard continues, calling on us to get off "this crazy treadmill." Naomi Klein puts it in a similar fashion in an article in *The Nation*: "We weren't born having to shop this much, and we have, in our recent past, been just as happy (in many cases happier) consuming far less."

This argument, that material advances can't bring you happiness, is a common, even ancient one. Money can't buy me love, etc. But it turns out it's not true at all.

The contemporary, ostensibly more evidence-based version of the happiness argument finds its origins in a classic 1974 sociological study, "Does Economic Growth Improve the Human Lot? Some Empirical Evidence" by economist Richard Easterlin. Using as basic data statements by subjects on their subjective happiness, he found that once people have enough wealth to meet their basic needs, higher levels of consumption do not correlate with higher levels of happiness.

The New Economics Foundation, a UK-based progressive economics think-tank, embraces Easterlin's work in their 2010 report, *Growth Isn't Possible*[43]:

People tend to adapt relatively quickly to improvements in their material standard of living, and soon return to their prior level of life satisfaction. This is known as becoming trapped on the 'hedonic treadmill', whereby ever higher levels of consumption are sought in the belief that they will lead to a better life, whilst simultaneously changing expectations leave people in effect having to 'run faster', consuming more, merely to stand still.

But what came to be known as the Easterlin Paradox has since been criticised for its methodology of using notoriously unreliable self-reported statements about a very subjective phenomenon such as happiness. Since that time however, an explosion in opinion surveys has allowed research in happiness studies (yes, it's a serious discipline) to base itself on a more rigorous footing, and a growing body of literature suggests that the opposite is the case: money really can buy you happiness.

In 2003, Dutch sociologist Ruud Veenhoven, a founder of the *Journal of Happiness Studies,* and his colleague Michael Hegarty published a fresh analysis using more extensive data sources (improving statistical power by using longer time series, by including nine nations with low GDP per capita and by pooling countries into income tiers), finding that in fact the paradox did not exist. Countries do become happier with increasing income. In an update to the study three years later, the authors found:

Average happiness has increased slightly in rich nations and considerably in the few poor nations for which data are available. Since longevity has also increased, the average number of happy life years has increased at an unprecedented rate since the 1950s.

That is to say that there are certain things that money can't buy of course, and the fruits of economic growth are nowhere spread evenly amongst citizens, but economic growth does allow for fresh investment in medical and scientific research that extends life and improves the health of those lives—a clear increase in happiness. If one is dead, manifestly, one cannot be happy.

Then in 2008, a pair of economists from the Brookings Institution, a centrist US think-tank, reached similar conclusions in their own exploration of opinion survey data. The worst that could be said was that happiness increases logarithmically—that is, after a big jump in happiness attributable to emerging from the depths of poverty, happiness then increases more slowly than income does, but at no point do happiness increases cease.

Extending the progressive nature of this analysis, we can say that economic growth allied to egalitarian protections shares out these advances to more people. It also permits less drudgery, more leisure time, and more opportunity to travel and visit friends and family.

There is a second classic study anti-consumerism advocates regularly cite, a 1978 paper, "Lottery Winners and Accident Victims: Is Happiness Relative?" by sociologists Philip Brickman, Dan Coates and Ronnie Janoff-Bulman in which they said that they had found that both paraplegics and those who struck it rich from the lottery initially experienced significantly affected happiness levels, but after a period of time, subjects habituated themselves to their new circumstances, with happiness levels typically returning to average levels. The authors described this phenomenon as the 'hedonic treadmill'.

However, again, this was an interview-based study, and not longitudinal—that is, it did not follow subjects over long periods of time. Since then, a number of longitudinal studies have found the opposite to be the case, that indeed happiness levels can vary drastically, particularly in cases of severe, long-term disease or disability, or in situations of imprisonment. Such studies have forced many researchers to conclude that the concept of the hedonic treadmill is unsustainable.

*

If we think about it, all this should be obvious. Being poor delimits your life profoundly, cranks up stress and anxiety, induces sleeping disorders, and ultimately provokes a flood of diseases such as high blood pressure, heart disease, diabetes, circumscribing your degrees of freedom in every way. The partisans of degrowth and anti-consumerism will say that of course, we mean after basic needs are met, everything else is a luxury. But what are basic needs? Strictly speaking, if we just think of ourselves as animals, all in essence that humans need to maintain homeostasis (the optimum, stable conditions for survival) is sufficient clothing and shelter to maintain an average core temperature of 37 degrees Celsius, around 2500 calories per day for men and 2000 calories for women, and about 1.6 litres of fluid per day for women and 2 litres for men. But even in most jails, prisoners receive more than this. Indeed, if they do not, we consider such prisons to be breaching the human rights of inmates. While we are indeed animals, we are also so much more than that, and unlike the rest of nature, our 'needs' extend well beyond the maintenance of homeostasis; they extend outward indefinitely. We are *always* working to improve our condition, to increase the degrees of freedom available to each of us. To say that we are always working to increase our *wealth* seems to be crude or greedy, but upon inspection, it is precisely the same thing.

We can see this easily by asking such questions as: Is electric lighting so all school pupils can study at night a basic need or 'stuff', a luxury? How about refrigerators that reduce the labour (usually women's labour at that) previously spent having to purchase perishable food daily? Running water, central heating, electric irons? What about the washing machine, the device left-wing economist Ha Joon Chang describes as a more important invention than the internet?[44] Chang writes that:

The washing machine, piped gas, running water and all these

mundane household technologies enabled women to enter the labour market, which then meant that they had fewer children, had them later, invested more in each of them, especially female children. That changed their bargaining positions within the household and in wider society, giving women votes and endless changes. It has transformed the way we live.[45]

How about indoor toilets? Are they a luxury? It's true that they once were. My mum tells me stories of running to her Nan's chilly outhouse in 1950s Walsall. Some bright young things who do not remember this time do seem to think that toilets are unnecessary. Leftist writer Yasmin Nair wrote a gloriously cantankerous essay in 2013 excoriating how "a number of factions believe that we have gotten *too* far from shit. They believe we are neglecting 'more natural' ways of dealing with shit."

The very idea that one should be concerned about privacy or dignity while shitting is one that hippy-radicals and academics mock. Historians of shit like to describe this urge to privacy as a bad departure for the culture. Some describe, with a smugness that sometimes floats to the surface, the fact that the ritual of going to the toilet was once upon a time a communal experience, with people laughing and chatting away and catching up on gossip as they went about their business. Dominique Laporte's 1978 *History of Shit*, for instance, interrogated shit as a reservoir of cultural anxieties and asserted that the development of modern toilets was symbolic of a larger bourgeoisification of culture.

...Get a bucket, we're admonished, some sand, and turn it all over every now and then and voila, you're taking care of the environment. The flush toilet has transformed lives for millions but it continues to be derided as a wasteful, almost evil, part of modern life. The charges seem unfair to a portal that makes lives easier for so many—there is nothing like a temporarily dysfunctional one to remind you of the necessary part it plays in life.[46]

Nair herself was born in Calcutta, before it became Kolkata, and she is having none of this shit. "Shit is literally everywhere," she writes about shining India, "and when I was growing up it wasn't uncommon to have to make one's way through the evidence of bodies having shat nearby or, for that matter, actually shitting near one because there was simply no place for privacy or dignity."

According to the UN, as of 2014, some 2.5 billion people in the world still do not have access to decent sanitation, and 1 billion have no choice but to defecate in the open, a practice that extends exposure to cholera, typhoid, hepatitis and polio, and puts women and girls at greater risk of rape and abuse. Every day, people in such areas drink water laced with somebody's faeces. As a result, diarrhoea is the third biggest killer of children under five in sub-Saharan Africa, and every year, 44 million pregnant women are infected with worms.[47]

But toilets and washing machines are too easy. So I'll go further. Is the Large Hadron Collider a basic need, palliating but not quenching our species' thirst for knowledge? How about the Mars Curiosity Rover, its name itself expressing that most human and unbounded of characteristics? Or are they just 'stuff'? One day, we may want to build a particle accelerator much larger than the LHC's 27 kilometres in circumference across the French and Swiss border. The particle physics community calls it their ultimate dream machine[48], the Fermitron, a particle accelerator spanning the circumference of the Earth or existing in a stable orbit around it. It will require truly mammoth wealth—economic growth—to build it. Again, is this a need or is it just stuff?

Rather, 'basic needs' and 'stuff' are not really useful

categories. Likewise the distinction between needs and wants is a false one. Instead, we should ask whether we have decided on our production priorities democratically, whether what we decide to produce is shared in an egalitarian fashion amongst all citizens, and whether this production maintains or diminishes the optimum planetary living conditions for humans. Over the course of consideration of these three questions, there isn't a clear and permanent upper boundary of needs. Instead, the maximum amount of what should be produced is always dynamically contingent on the political economy and level of technology of the particular time.

Do we at least want to do away with planned obsolescence, as Leonard and Klein demand? Of course this is an utterly irrational waste of resources, but such a trick of industrial design is woven into the very fabric of capitalism, of capital's requirement of self-valorisation. Ridding ourselves of it is not going to happen by consuming less, but by establishing democratic control of the economy—by rational planning. It is the market that drives planned obsolescence, not growth or consumerism. True economic democracy is a rather grand ambition of course, but once again we can see the poverty of ambition, the accommodation to capitalism of these alleged anticapitalists.

*

To be clear, none of my argument serves to buttress the rightly maligned trickle-down myth—that a rising tide lifts all boats. In the absence of a distributionist, egalitarian paradigm, economic growth tends to advance the welfare of the few, rather than the many. Growth in inequality is an inescapable product of capitalism, as underscored in voluminously evidentiary detail by French economist Thomas Picketty. Nor does my argument require an endorsement of Gross Domestic Product (GDP) as a useful metric of the wealth of a society. Many of the critiques of GDP are perfectly sound. GDP doesn't take into account non-market interactions, the underground economy, subsistence production, and product-quality improvements. And, famously, GDP counts the clean-up performed repairing the damage done by Hurricanes Katrina and Sandy and the BP oil spill in the Gulf of Mexico as adding tens of billions to America's GDP. But there are other, better metrics that have been proposed such as the Human Development Index (HDI), the Inclusive Wealth Index (IWI), and the Index of Sustainable Economic Welfare (ISEW). Some of them are better than others. The Happy Planet Index (HPI) introduced by the New Economics Foundation and championed by Friends of the Earth switches the drawbacks of GDP for its own defects, particularly the use of the spurious, politics-of-limits notion of 'ecological footprint', which is all but a synonym for carrying capacity. The Worldwide Fund for Nature (WWF) annually issues a thick, typically 180-page Living Planet Report, with its own, trademarked Living Planet Index (LPI) that also embraces the concept of ecological footprint and uses an updated version of the model employed by the Club of Rome for its *Limits to Growth* report in the 1970s. Its 2014 edition tells us that as a result of 'overshoot' (there's that ejaculatory word again!), 'we' would need the regenerative capacity of 1.5 Earths to provide the ecological services we currently use. Yet if you just snip out energy from the total footprint (a reasonable thing to do, as we can replace fossil fuels with cleaner sources in a snap, given the political will), then the ecological footprint per person has been *declining* since 1960.[49] And any new metric that can be cheerfully embraced by UK Tory Prime Minister David Cameron[50] as a possible replacement for GDP, as prior to taking office the Conservatives did, issuing a report that welcomed the HPI, should be given the fiercest of side-eyes. But the debate over what should replace GDP could fill another book. The salient point here is that agreeing that GDP is an utterly perverse metric does not require simultaneously embracing an anti-consumerist or degrowth perspective. Separate out the critique of GDP from the critique of growth and you'll have me cheering you on.

Finally, none of this should be seen as an argument against ever more efficient systems of recycling, even of composting and the re-use of human waste (although the emphasis in the latter two cases should always first be upon avoidance of increased risks to public health, which too many current proposals along these lines blithely ignore—more on this in a few chapters). To achieve ever-expanding economic growth without inadvertent damage to ecological services that benefit humans requires greater and greater efficiencies in the use of material resources. Recycling, done sensibly, can play a large part in achieving these efficiencies, as well as more widespread adoption of industrial and urban systems that conserve materials. It would be absurd to be pro-waste, to be pro-profligacy. But the biggest achievements to this end will come from an emphasis on technological trans-formation in infrastructure and at the point of production—usually the focus of progressives—rather than on the choices of individual consumers—which are usually the focus of more neoliberal forces. To take just one example, but a crucially important one, construction and demolition represents one of the largest contributors by mass to the waste stream. There is very little indeed that can change here as a result of consumer choices. Instead, change that would achieve significant results would require different building strategies and technologies in new construction.

And according to historical data, we find that after a certain level of wealth, the amount of rubbish produced per capita tends to level off at about 1.4 kg of solid waste per day. (And Japan produces about a third less waste per capita than the US, largely as a result of much higher-density city living—in other words, more modernity, more urbanism, the opposite of the back-to-the-land approach of the janissaries of anti-consumerism.) The bulk of the expected growth in waste over the rest of the 21st Century before peaking around 2100 will come from the Global South as they develop, which they of course are entitled to do.

We cannot deny them their right to trash.[51]

*

I became a shouty socialist teenager by way of the church, or more accurately, via my personal readings of scripture and what I later realised were left-wing priests. There were other factors, but the simple life of Jesus the carpenter's son; His Sermon on the Mount; the writings of liberation theologians such as Gustavo Gutierrez and Leonardo Boff; the example of the revolutionary Jesuits in the FSLN and the assassinated archbishop of San Salvador, Oscar Romero; and, ahem, somewhat more cringe-worthily, Franco Zeffirelli's sentimentally overripe film *Brother Sun, Sister Moon*—about the spoilt young nobleman who gives up all that he owns to live a life of poverty caring for lepers, later canonised as Saint Francis of Assisi—were all seminal influences. I'm an atheist now and have been for years, yet thinking on such characters and their ethics moves me still. "It is easier for a camel to go through the eye of a needle than for a rich man to enter into the kingdom of God," (Matt. 19:24); "So the last shall be first, and the first last," (Matt. 20:16) etc., etc. Such proto-communist Biblical verses still trip off my tongue as easily as Sunday school jokes about Jesus biting his nails. And to my mind, the anti-consumptionist mindset results from a fundamental confusion between steady improvements in societal wealth, which we should indeed all desire, and individual greed, which has correctly been denounced by so many prophets, playwrights, poets and jesters since the beginnings of class society. The fury expressed by these early heroes of mine at the rich who strive for property at the expense of others is a legitimate and righteous hatred. Yet a desire for all of society to advance together, ever forward without limit—to expand the commonweal—is not an expression of greed at all, but the noblest of yearnings. Though to this day I can appreciate the spiritual aesthetic and act of solidarity that is the simple life (indeed, as a working journalist, I don't think I'll ever manage to escape the simple life myself), anti-consumerism is not, in Gutierrez's words, a "preferential option for the poor."

Inequality should not be replaced by an equality of poverty, but an equality of abundance. As the British suffragette Sylvia Pankhurst clarified succinctly on the front page of the *Workers' Dreadnought* newspaper in 1923, arguing against the levelling-down egalitarianism popular then as now amongst some hair-shirted tendencies of left-wing thought:

> Socialism means plenty for all. We do not preach a gospel of want and scarcity, but of abundance. Our desire is not to make poor those who to-day are rich, in order to put the poor in the place where the rich now are. Our desire is not to pull down the present rulers to put other rulers in their places. We wish to abolish poverty and to provide abundance for all. We do not call for limitation of births, for penurious thrift, and self-denial. We call for a great production that will supply all, and more than all the people can consume ... To-day production is artificially checked, consumption still more so.

What have we been doing lo these many decades as trade unionists, if not fighting to steadily advance the prosperity of our class?

A socialist never stops demanding more.

6

Locally-Woven Organic Carrot-Pants

Wait, wait, wait! Doesn't Naomi Klein, distinct from some of these other degrowth militants, call for exactly what I have been arguing is necessary, a return of planning? In fact, doesn't she say: "A serious response to the climate threat involves recovering an art that has been relentlessly vilified during these decades of market fundamentalism: planning. Lots and lots of planning." And doesn't she add, in an interview with the *Indypendent*, a radical newspaper out of New York: "It's not about, 'Hey, let's start an energy co-op.' No. That kind of fetish for very small-scale initiatives won't get us where we need to go,"

She does say all of this, but it's all quite a muddle, and at other times she argues the opposite. At best, this sort of thinking offers very little coherence, and at worst unconsciously buttresses neoliberal assaults on the public sector.

Klein makes no distinction between *democratic* planning, as in the case of the UK's very-big-indeed National Health Service (the fifth largest employer in the world[52]), and *authoritarian* planning, as in the case of the USSR or Walmart. (Yes, Virginia, while Walmart, the third largest employer in the world, operates within the free market competing against other shops, internally, the multinational firm is the very model of planning, as are all firms. Highly hierarchical and, yes, dictatorial, but planned with brilliant efficiency by humans nonetheless. As American Marxist literary critic Fredric Jameson has scandalously suggested, strip out the exploitation of its workers and the lack of democracy, and the stunning logistical wonder that is Walmart actually becomes an example of planning that socialists should study with keen scrutiny. Walmart is, Jameson asserts cheekily but with sincere admiration, "the shape of a utopian future looming through the mist, which we must seize as an opportunity to exercise the utopian imagination more fully, rather than an occasion for moralizing judgments or regressive nostalgia.")

Rather, Klein distinguishes between types of planning based on *scale*. Contra her comment in the *Indypendent*, she most often argues that the good type of planning is small, local and non-statist, and the bad type is big, transnational and *statist*. Just over a fortnight after that interview, she told *Roar Magazine* that small-scale local energy co-ops are absolutely the solution! "I think another inspiring movement is the rise of renewable energy in Germany," she said. "But what's really interesting about it, is that it is the small-scale, decentralized, cooperatively-owned aspect of the transition that is fastest-spreading, that has people most excited."[53]

Klein's fear of the big-and-international and fetishisation of the small-and-local runs through most green-activist energy proposals, which regularly call for decentralised, renewable energy production rather than 'Big Kit' projects. To the great glee of neoliberal ideologues, the green Mr Magoos not only blithely ignore the energy-market liberalisation upon which such independent power producers (IPPs) depend, but actively give these vultures the eco-friendly fig leaf they need to push through deregulation, unbundling and privatisation in the face of opposition from public-sector unions and consumers.

Similarly, much like typical red-baiting assaults on the idea of socialism from the right, Klein attacks "the statist left," conflating the bureaucratic-kleptocratic regimes of the USSR and Mao's China with modern proposals for *democratic* planning. She talks of "the failures of industrial state socialism" and says that "the reality is that Soviet-era state socialism was a disaster for the climate. It devoured resources with as much enthusiasm as capitalism, and spewed waste just as recklessly … China's *command-and-control* economy continues to be harnessed to wage an all-out war with nature, through massively disruptive mega-dams, superhighways and extraction-based energy projects, particularly coal."

It is as if Klein and her co-thinkers are unfamiliar with the 90-odd years of socialist but anti-Stalinist discourse from the likes of Rosa Luxemburg, Anton Pannekoek, Leon Trotsky, CLR James and their intellectual descendants, who distinguished between genuinely democratic planning on the one hand and central planning set by bureaucrats divorced from democratic control on the other. But Klein accepts at face-value China's claim to socialism, and so for her the main problem with the PRC seems to be its bigness, rather than the fact that it isn't democratic.

There is a difficult discussion that the left needs to have about our attitude toward the state. Nationalisation is not synonymous with socialism. As Friedrich Engels wrote about state ownership: "Otherwise, the Royal Maritime Company, the Royal porcelain manufacture, and even the regimental tailor of the army would also be socialistic institutions, or even, as was seriously proposed by a sly dog in Frederick William III's reign, the taking over by the State of the brothels." But such a developed, nuanced, grown-up discussion about the public sector is not what is happening here at all.

Writing in *The Nation*, Klein says she wants to see more subways, street-cars and light-rail transit systems, and smart electrical grids carrying renewable energy. She is absolutely correct to say that the private sector is "ill suited to providing most of these services because they require large up-front investments ... some very well may not be profitable. They are, however, decidedly in the public interest, which is why they should come from the public sector."[54] But most of these projects seem pretty large-scale and 'statist left' to me. Subway trains and street cars can't be knitted together by skinny-jeaned urban apiculturalists. Metal has to be mined and plastic parts manufactured by industry. And a wholesale shift of transport from being fossil-fuel-based to running on clean sources of electric power will require a massive ramping up of electricity production, and, as Klein acknowledges, smart, load-balancing electrical grids. How is this not Big Kit?

As part of her plan to save the planet, Klein also calls for a guaranteed basic income: "A basic income that discourages shitty work (and wasteful consumption) would also have the benefit of providing much-needed economic security in the front-line communities that are being asked to sacrifice their health so that oil companies can refine tar sands oil or gas companies can drill another fracking well."

One can debate the merits of this social-democratic idea and whether a drip-feed of income from the state is really a progressive stance compared to a strategy of working to revive the strength of the labour movement so that working people ourselves independent of the state steadily advance a claim on the wealth that we produce for our bosses (I plainly favour the latter over the former, but the advisability of basic incomes is a debate for another time). Regardless of one's opinion on the matter of a guaranteed basic income, it would, of necessity, *boost* consumption, contradicting Klein's anti-consumption stance. If more people have more money, they will buy more things. Furthermore, a basic income would involve planning on a national, not localist, scale, which Klein even acknowledges a few paragraphs later.

But how exactly are any of these sorts of grand, social-democratic policies and public-sector infrastructure projects to be organised? Is such coordination "command-and-control" or is it democratic planning?

*

The error Klein makes is the central error most of the anti-growth theorists make: for all their fulmination at corporations, profit and greed, they have not made any serious study of the internal logical contradictions of the capitalist system, or of the state, of how the machine works. Klein makes no attempt at explaining why it is that capitalism is bad for the environment other than that, like The Blob, 'it grows'. It is remarkable for a book with the strapline "Capitalism vs the Climate" that Naomi Klein's *This Changes Everything* doesn't even define what capitalism is. Instead, 'capitalism' for Klein and friends is just sort of assumed to be something big and mean that grows and seeks profits.

So okay, capitalism does seek profits, and it must keep growing or it stalls, but we've had over a century and a half of careful, evidence-based, in-depth heterodox analysis of capitalism by critical economists and sociologists describing its mechanisms and contradictions, explaining how it so much more than that. Yet she doesn't feel it necessary to even thumb through such scholarship. She regularly refers not to capitalism as a mode of production as the problem, but "deregulated capitalism" or "free-market fundamentalism." Now, I'm certainly all for a critique of the deregulatory mania we've experienced since the 1980s, but does Klein mean to say that if capitalism were simply regulated better, or the free-market weren't so "fundamentalist," it would be fine? This contradicts what she says elsewhere that capitalism is a problem because it is dependent on growth. But how exactly do you regulate away such dependence? It's like regulating sugar cubes so that they stop containing sugar.

And this in essence is the same error for which Friedrich Engels criticised early social reformers or philanthropists such as Count Claude Henri de Rouvroy de Saint-Simon, Charles Fourier and Robert Owen, who produced their own idiosyncratic visions of the ideal civilisation. Such utopian socialists were, Engels declaimed, capable of well-meaning moral and idealistic indignation at injustice and misery, but incapable of attempting a properly materialist, systemic analysis of how capitalism produces this. And without such thoroughgoing and rigorous—as Engels described it, 'scientific'—analysis, there was no guarantee that their hypothetical visions of the perfect organisation of society could escape capitalism's contradictions: "These new social systems were foredoomed as utopian; the more completely they were worked out in detail, the more they could not avoid drifting off into pure phantasies."[55]

Klein accepts without contest the basic liberal critique of socialism. The founder of the left-wing *Jacobin* magazine, Bhaskar Sunkara, describes Klein as the archetypical "anarcho-liberal," noting that she had said during a New York panel discussion that she was opposed to "grand projects of human freedom" and that a *New Yorker* profile described her philosophy as "distrust[ing] centralization, institutions, platforms, theories—anything except extremely small, local, ad-hoc, spontaneous initiatives."

Sunkara's prognosis is most apropos when it comes to Klein's contradictory stance of simultaneous localist opposition to the state while demanding public-sector transport and energy projects, of wanting planning but not really thinking through what planning means: "The worst of both worlds, the 'anarcho-liberal' can neither manage the capitalist state nor overcome it, and aspires to do both and neither at the same time."

I'll happily concede that there are many, many times when Klein makes excellent arguments that are, essentially left social-democratic critiques of corporate environmentalism, the 'Blairite' ecologists, if you will: against emissions trading and carbon offsets; for the reversal of privatisation; for tough regulation of businesses; for higher taxes on the wealthy; about how fossil-fuel lobbyists work their regulatory-capture magic; how investor-rights language in trade agreements helps unravel environmental regulations; uncovering the rightward, industry-collaborationist drift of the big green NGOs; and why we should be highly sceptical about the motives of 'green capitalists' such as Starbucks and Chipotle. (Indeed I spent much of the 2000s as an environmental journalist in the capital of the European Union exposing corporate lobbying, the failures of the EU Emissions Trading Scheme, Europe's bullying climate diplomacy, the neo-colonialism and ineffectiveness of biofuels and carbon offsets, and green groups' support for energy liberalisation and privatisation.)

Yet one can certainly agree with all of this and still not concur that size or growth is a problem. The *primary* attribute that we seek with any climate and energy project is of course how much of a contribution on aggregate it makes to the decarbonisation of our economy. Simultaneously, we demand that such endeavours be implemented democratically. Size or scale should not be factors. Small but undemocratic is not superior to big and democratic. If democracy has an upper bound, then small-is-beautiful theorists will have to explain the mechanism through which this upper bound imposes itself, and whether 'big' means geographically big or demographically big or financially big or what. Is Canada, the second-largest country in the world, less democratic than the tiny autocracy of Bahrain? Is democracy simply infeasible in places like China or India? Is even internationalism then inherently undemocratic due to its maximally ambitious scale?

This small-is-beautiful localism is a remarkably pessimistic position, and certainly at odds with the long-cherished, optimistic goal of socialists of *global* economic democracy.

*

We need a "new civilizational paradigm," says Klein, one that will require a systematic devolution of power to the local level "whether through community-controlled renewable energy, local organic agriculture or transit systems genuinely accountable to their users."[56] We must permanently park our "cargo ships, jumbo jets and heavy trucks that haul raw resources and finished products across the globe devour[ing] fossil fuels and spew[ing] greenhouse gases," she says, in order to relocalise production.

Any long-haul transport would need to be rationed. In a 2011 piece for *Rabble*, a Canadian leftish online outfit, Klein counterposes "big, centralized technologies that confirm their belief that humans can dominate nature" to the locally-woven organic carrot-pants and peak-oil millenarianism of the Transition Town movement.[57] This is the international network of grassroots activists readying themselves and their mid-sized market towns with roof-top windmills and alternative currencies and "energy descent action plans"[58] for what they believe to be the coming energy crash, a mirror-image of the right's Christian Dominionists and survivalists certain of the imminent arrival and cleansing of the Rapture and/or race wars.

The New York Times describes this multi-billion-dollar-a-year sector the 'Mad Max Economy' (both Costco and Walmart for example now offer a year's supply of food freeze-dried and in cans).[59] But perhaps 'mirror-image' is even the wrong term, as anyone investigating the subcultures of self-sufficiency, modern homesteading[60] and disaster 'preppers', with their hunting workshops, home-canning seminars, DIY water-filtering projects and camp-craft weekends, will notice how at no point do you find a clean break between the deep ecological catastrophists awaiting the Peak Oil reckoning and the happy-clappy hard-right god-botherers or 9/11 Truthers readying themselves for final conflict with the New World Order. They all tend to speak in a similar, catastrophist dialect. The folksy term 'preppers' comes from their hobby of *preparing* for the end of civilisation. *Doomsday Preppers*, a National Geographic Channel breakout hit, has been running for four seasons now.[61] Tony Tangalos, the host of *The Prepper Patch*, a survivalist talk show, is in the planning stages of establishing "Christian Transition Villages" in the Wikieup area in northern Arizona "for people who believe that we may be close to experiencing some dramatic man made or naturally occurring changes in our world, 'Black Swan Events'; can move to and begin living the agrarian, self-reliant lifestyle, totally 'off the grid', with solar power, well water, on site crops & livestock." Tony and his fellow transitioners "are seeking additional people who share our concerns for the need to transition from the urban lifestyle to a more rural, agrarian lifestyle where we provide for most all of our needs from within our sustainable village ... Able bodied applicants with the ability to demonstrate experience in needed skill sets such as farming, ranching, blacksmithing, security, medical training, physical fitness training and leadership will receive preference."[62] (Although, as far as I know, there are no scything workshops.)

Is it fair to make this comparison between peakist Transitioners and survivalist Preppers? The Transition response to climate change and what participants believe to be Peak Oil and resource depletion, "the issues that will increasingly dominate our world," in the words of one transitioner, involves local food initiatives such as community supported agriculture (CSAs), urban food production, garden shares, promoting cycling, tree planting, developing community renewable energy companies and establishing alternative local currencies. They teach each other practical skills such as building straw bale houses and cob huts with local materials and generating their own energy in order to improve their community's sustainability and resilience in the face of the shocks to the system that they are certain are coming soon. The concept has spread incredibly fast. There are now 1130 registered Transition Towns in 43 countries as of writing, having exploded out of the Ground Zero of transitiondom, the 8,000-soul town of Totnes in 2006, the brainchild of the movement's chief guru, permaculturalist Rob Hopkins. It's not merely the hippie fringe in these communities either: counties, boroughs and even cities have officially declared themselves to be Transition towns.

"In the cities and towns that have taken this responsibility seriously, the process has opened rare spaces for participatory democracy," says Klein, "with neighbours packing consultation meetings at city halls to share ideas about how to reorganize their communities to lower emissions and build in resilience for tough times ahead."

In a 2009 TED talk in Oxford, Hopkins explained that participants say they find developing projects readying themselves for the end of the "petroleum interval" and designing *energy descent action plans* a really joyful, playful, creative, exciting experience.

It sounds lovely, doesn't it? What progressive could be opposed to building community and participatory democracy?

It turns out though that the core of the sort of projects that Transitioners get up to, localising agriculture, doesn't drastically reduce carbon emissions, but rather increases them, as I'll demonstrate more extensively shortly (as local agriculture boosterism has taken off far beyond the confines of Transitionism in recent years).

On the matter of local 'complementary currencies', we find here too that this may actually result in greater greenhouse gas emissions for the simple reason that the sort of larger loads in your car when you travel to the non-local-currency-accepting big box retailer can mean fewer trips there. In addition, the big box stores themselves can have fewer deliveries than corner shops, resulting in fewer emissions per hammer or shower curtain or jumbo bag of kitty litter. In any case, is that hammer, shower curtain or kitty litter made locally anyway? Even local traders have transnational inventories. Big box stores are often cheaper than mom-and-pop outfits, making local currencies an encumbrance to impoverished families. As for 'keeping money circulating locally', the aim here is, one would assume, plugging holes that leak wealth out of a community. But money is not itself wealth; money is a method of keeping accounts. Typically, local currency schemes are one-to-one convertible with normal pounds or dollars, so they're just swapping one system of accounting for another. Improving the wealth of a community requires economic development, something at which the

Transitioners look askance. And sociological studies investigating the viability of alternative currencies suggest that they usually collapse after a few years, but work best in middle-class left-liberal communities—that is, those regions already pretty well off.

Thirdly, as a cyclist myself for whom near-death experiences resulting from stone-blind drivers are as much of a morning ritual as coffee and snooze-button-molesting, I fully support the greater safety that will come from properly separated bicycle lanes as exist in cities like Amsterdam and Copenhagen. I even sort of think that once a year, every motorist should be required to cycle end-to-end through a major metropolis at rush hour just once before having their driver's licence renewed. But this is so very, very much inadequate to the gargantuan scale of the problem of climate change. We need an electrification of the global transport fleet, which will require vast increases in electricity production, not everyone shifting to bikes. This is not going to come from toy energy co-ops. And the whole Transition worldview promotes an austere politics-of-limits mindset that retards human flourishing. We are confronted with a desperate international housing crisis in the West, in which countries are millions of homes short of the necessary replacement of dilapidated housing stock, never mind building new homes to keep up with population increases, partially as a result of a raft of NIMBY-ish conservation-inspired building restrictions against new build and densification. Yet Hopkins describes energy descent plans as a local 'Plan B' that assumes a time when there won't be more energy, more jobs, more homes next year: "What does it look like if we actually embrace that?" I'm sick of paying ridiculous sums for my rent. I have no desire to embrace even fewer homes.

Finally, Labour leader Ed Miliband may have been a 'keynote listener' at a Totnes conference on the concept, but it's the Conservative Party that has most warmed to Transition Towns, as Rob Hopkins at one point is quite clear: "[P]olitically, the thing about that that's interesting is that actually the question of relocalisation is much more familiar territory for the right than for the left ... Certainly at the moment with [Prime Minister David Cameron's] Conservative agenda it's very much around localism and devolving power to local councils and all that kind of stuff – at the same time as cutting money from everywhere. But actually, that idea of supporting local economies is an idea that sits much more comfortably with the right than with the left. Certainly in terms of Transition, there's more familiar territory, in some aspects of what we're trying to do, on the right than on the left, which has been interesting."[63]

In the end, we have to say that the best Transitioners can offer is an exercise in effective community building. But the shotguns 'n' whiskey aroma of the Preppers shouldn't distract us from the fact that they're building community in their own way as well. We laugh at the survivalists because we know that their reasons for going off grid have been bogus, from Y2K to the Rapture to the New World Order. Yet a lot of the Transitioners' ideas and methods seem pretty bogus too. If their community building efforts can be put to more useful goals than localism, then that of course would be great. But it would be also great if survivalists' community building could be put to more useful goals than biblical eschatology.

Even the Jonestown Peoples' Temple Agricultural Project *built community.*

*

But the Kool-Aid of the cult of localism is not just being drunk by Rob and Tony and Naomi. Localism is pushed by Bill McKibben—the ex-*New Yorker* journalist, initiator of the 400,000-strong People's Climate March outside the UN climate talks in New York in the fall of 2014, and supremo of international climate-change activist group 350.org—in his latest book, *Eaarth:*

Making a Life on a Tough New Planet (yes, that's spelt correctly—
McKibben added an extra 'a'). Localism is the focus of novelist
Barbara Kingsolver's *Animal, Vegetable, Miracle,* a memoir of her
family's efforts to eat only food that they had grown themselves
or obtain locally for a full year; as it is of *The 100-Mile Diet* by
Alisa Smith and James MacKinnon, and most of food writer
Michael Pollan's oeuvre. There's *Local: The New Face of Food and
Farming in America* by Douglas Gayeton; *The Locavore's Handbook*
by Leda Meredith and Sandor Ellix Katz; cookbooks like *Local
Flavors: Cooking and Eating from America's Farmers' Markets* by
Deborah Madison. Twee little signs hand-calligraphed or rubber-
stamp-printed on moss-green parchment and lavender-blush
vellum card-stock in cafes, farmers' markets and high-end
grocery stores declare the localist virtue and upstandingness of
their muffins, cranberry horseradish and herbal alternatives to
deodorant. Busybody Facebook commissars enforce localist
doctrine criticising the consumer choices of their friends (when
they're not judging their parenting choices). The local food
movement has achieved such ubiquity that it became the
mocking subject of satirical comedy series *Portlandia* in a sketch
called 'Is it Local?', in which a pair of ethical restaurant-goers
grill their waitress about the sustainable pedigree of the dish they
are thinking of ordering, which involves a woodland-raised,
heritage-breed chicken that has been fed a diet of sheep's milk,
soy and hazelnuts, from 30 miles south of Portland, and is named
Colin.

It seems so simple: food (or anything else) produced locally
will not require the carbon-spewing transportation of such items
via cargo ship or truck or plane from far away. It appears to be an
easy rule of thumb enabling consumers to do the right thing.

But the reality is a great deal more complicated. Instead of the
crude heuristic of 'food miles', if we are genuinely concerned
about greenhouse gas emissions, we need to make sure we are
actually doing good, not just feeling good. That means that we

need to base such decisions on full life-cycle assessment (LCA) studies—a method of analysis that takes into account all aspects of the production and distribution of a product.

And when we do look at LCAs, for some products, it turns out that yes, indeed, it does make sense to relocalise production, but for many, many other items, the economies of scale involved make the amount of energy employed and thus greenhouse-gas emissions per item far less than an item that is locally produced, despite the thousands of 'food-miles'. According to a 2005 UK Department of Environment, Food and Rural Affairs analysis,[64] tomato farmers in sunny Spain produce less CO_2 than tomato farmers in frequently overcast Britain employing heated greenhouses (630 kg of CO_2 vs 2,394 kg of CO_2 per tonne). The same is true of Kenyan versus Dutch rose growers, with the former producing six tonnes of CO_2 per 12,000 roses cut, and the latter producing 35 tonnes of CO_2 for the same amount.[65] It is the *production* of food that that has the largest energy appetite, rather than transportation. Again, it is simply more efficient to have the roses grown where flower production depends almost entirely on the warmth of the sun in equatorial Kenya rather than on the heating and lighting systems of the temperate Netherlands.

A similar investigation in 2008 by Carnegie Mellon researchers Christopher Weber and Scott Matthews,[66] covering the American situation, found that 83 percent of an average household's carbon footprint came from emissions during the production phase, with just four percent of full life-cycle greenhouse gas emissions coming from transport from producer to retailer. Weber and Matthews found that due to the different *carbon-intensity* of the production and distribution of different items, with red meat on average roughly 150 percent more carbon-intensive than chicken or fish, a far more effective rule of thumb than "buying local" would be a dietary shift away from beef and milk. "Shifting less than one day per week's worth of calories from red meat and dairy products to chicken, fish, eggs or a vegetable-based diet achieves a greater greenhouse gas reduction than *buying all locally sourced food*," they conclude.

In a similar fashion, in terms of the amount of water used, it can be far more sensible to produce food in areas with heavy precipitation than in arid zones, reducing the need for irrigation, disruption of natural river flows, and piercing of aquifers. Some 70 percent of our freshwater use occurs in agriculture, so this should be a key concern of the localist eco-defenders.

Geographer Pierre Desrochers and public policy analyst Hiroko Shimizu describe how agriculture that is local, small-scale, less-technology-intensive—and crucially, by definition, low in productivity—is necessarily more *extensive*, that is, it uses up *much more land* for the same amount of food. There is a very simple reason for this. Not every plot of land, with its particular climate, soil type, geology, topography and so on—its terroir, if you will (and I use that term fully aware of the irony of its presence in an essay arguing against localism)—is equally well suited to all types of plant and animal. Specialisation and a division of labour between different regions that are better at growing different items is thus a more efficient use of land: you'll get more calories produced per hectare.[67]

The inverse of this process—disintensification, which localism requires—means turning more forest, wetlands and grasslands into agricultural space, releasing vast quantities of carbon in the immediate term and, in the future, eliminating the carbon sinks that forests would have represented. This process of indirect land-use change is essentially why biofuels have proven to be no climate solution. The defenders of localism are in thus little different to the biofuels industry, clinging to a particular agricultural practice long after the evidence has shown it to actually exacerbate climate change.

A focus on local seasonality fails for the same reason. If we say: Buy as seasonally as possible, the first question that must be asked in response is: Which region's seasonality are we talking about? New Zealand's apple harvest season happens when it's winter in the UK, making it more sensible to ship fresh granny smiths all the way from the Antipodes to Europe than to keep British apples in cold storage for six months. The same goes for New Zealand lamb, dairy products and onions, according to a trio of researchers at Wellington's Lincoln University.[68] Meanwhile another 2003 study from German researchers Elmar Schlich and Ulla Fleissner[69] found via a full life-cycle assessment that large-scale Brazilian orange juice producers shipping their product around the world had lower per-unit energy demands than small-scale German apple juice squeezers driving their truck just ten kilometres to market. If the advice instead is not local seasonality, but 'global seasonality', picking things to eat when they're in season wherever they come from, then yes, in principle, you may see some carbon emission reductions due to shorter storage periods. But in the modern era, most food items are always in season *somewhere* in the world. This isn't true for all items, and for such products, a preference for their seasonality might make sense, but then again, this should be assessed on a *case-by-case basis*, using an LCA to take into account all the other variables related to carbon emissions. To do this would require something like a very detailed spreadsheet comparing all the different products and their component inputs, transport, storage requirements and packaging rather than the clumsy heuristic of "Buy seasonal!", which, as demonstrated, in a number of cases is actually detrimental in terms of mitigating climate change. Such Excel Hell might make sense for more rational agricultural planning, but as far as an individual consumer is concerned, it would be far more effective to expend one's time fighting for clean energy infrastructure than on this sort of faff.

*

We also regularly hear from local farmers' market boosters of the alleged benefits of less packaging. The German concept super-market Original Unverpakt (of Kreuzberg, Berlin) claims to offer *zero* packaging, as does In.Gredients (of Austin, Texas) and the recently closed Unpackaged (late of Dalston, London). But this is a *bad* thing for the climate, not a good thing. When looking at the full life-cycle assessment of a product, waste and spoilage have to be taken into account. Any food that is wasted represents utterly pointless carbon emissions, greenhouse gases released in the process of production that ultimately served no human need. Yet packaging protects against bacteria, worms and other pests and prolongs shelf life. Lack of packaging thus not only contributes to increased public health risks, it can reduce carbon emissions efficiency. The anti-packaging jihadis think that they are reducing waste, but, however counterintuitive it may be, the opposite is the case.

Similarly, we hear of how the variety of shape, size and imper-fection of products at farmers' markets demonstrate 'authen-ticity' and allows products to be sold that would otherwise be discarded. In reality, standardisation and product-grading speed up shipping as buyers can be sure of what type of commodity they are purchasing with fewer inspections, thus reducing transport emissions. Regular size and shape also permit more efficient packing, thus decreasing the per-unit carbon emissions. Standardisation and grades also allow the more rapid diversion of faulty produce unfit for human consumption to animal feed or industrial alcohol production—another efficiency.

Community-supported agriculture (CSA) schemes—in which consumers pledge to support a local farm at the beginning of the growing season in return for typically weekly deliveries of produce boxes of whatever happens to have been harvested—are another increasingly popular effort amongst localists. Consumers feel they are doing the planet a service while offering economic stability to small local farmers, who receive a guaranteed market for whatever they produce. Yet one of the main reasons listed in surveys of CSA subscribers for abandonment of such schemes is the sheer amount of waste involved when households regularly end up with more food than they know what to do with ("What on earth do I do with six turnips?"), compared to if they had simply gone to the supermarket and only bought what they needed, and end up tipping the veg and fruit into the compost bin or into the trash. The problem is so common that *The CSA Cookbook* by Linda Ly is subtitled: *No-Waste Recipes for Cooking Your Way Through a Community Supported Agriculture Box*. While there is a mountain of anecdotal evidence of the problem, there is as yet little comprehensive scholarly research into the true scale of CSA waste. But already, some 20 percent of US greenhouse gas emissions come from agriculture, so if it is shown that significantly larger amounts of food waste are typical of CSA schemes, then widespread adoption of them would be more than counterproductive. Should we be as judgemental about local CSA enthusiasts as we are about people who drive Hummers?

Support for farmers would be better provided through such options as public-sector- or producer-cooperative-controlled marketing boards like Canada's wheat pools and later its wheat board, which, until its recent privatisation, offered producers guaranteed prices and distributed income. Largely a product of Canada's history of prairie socialism (which had ideological analogues in Australia and the American Midwest that produced similar models, and which may have been agrarian-egalitarian but was at the same time resolutely modernist, embracing the latest farming techniques), the marketing board models offer farmers much longer-term economic stability through the swings of bounty and crop failure than CSA schemes, while centralising commodity storage and distributing to where there is demand, thus reducing waste. But such large-scale centralisation, mechanisation, continent-wide movement of commodities, and relative anonymity of producers with respect to consumers is the opposite of localist farming. Interestingly, conservative attempts to break up agricultural marketing boards, which they seem to place in the same category of abominations as gay marriage, the hijab and public radio, have depended upon propagandising complaints about such monopsony systems from alternative agriculture operators (farmers' markets in particular) who complain about what they feel are overly restrictive rules on how and to whom they can sell their products (rules that are nonetheless necessary to ensure all farmers economic security).[70] There are still some 80 marketing boards in Canada, despite the Tories' deep antipathy, including ones covering egg farming, to take one example. These boards control the number of eggs produced in order to maintain a steadier price in order to deliver a steady supply of eggs to consumers while ensuring farmers can make a decent living. But these are not the eggs that are in high demand from localist foodies, it's small, organic producers driving their fresh eggs to farmers' markets, according to Canadian media reports, who rage against the "egg police" crackdown on their free-riding off the marketing-board system. "It's a huge issue," Tom Henry, a Vancouver Island farmer and editor of the magazine *Small Farm Canada*, told Toronto's *Globe and Mail* newspaper. "The right to sell eggs is the small-farm equivalent of the right to bear arms."

The flourishing of the whole local and foodie scene is dependent on a policy background of liberalisation, deregulation and privatisation.

*

How do we even define what counts as 'local' or the boundaries of the regional 'foodshed'? Fifty kilometres away? What about 150? Is raw cocoa flown in from the Ivory Coast but manufactured into chocolate down the road considered local? Would one of the most famous books on the subject, *The 100-Mile Diet* by

Alisa Smith and J.B. MacKinnon, have a different limit to what gets to be counted as local if Americans used metric rather than imperial weights and measures? *The 160-Kilometre Diet* doesn't quite have the same ring to it.

At a recent public meeting on some random aspect of left-wingery that I attended, a woman got up to encourage audience members to join her and her neighbours' local food swap-and-sale scheme they had set up. All you had to do was when you had some item or service you wanted to offer, you posted it to the group's Facebook page, which you could instantly access via your mobile, making swaps fast and easy. I did not get up to ask whether the scheme was also locally manufacturing phones, but perhaps I should have.

Labour conditions also go frequently under-considered by locavores. Political scientist Margaret Grey spent ten years researching the small- and medium-sized family farms of the Hudson Valley in New York, writing up her explosive findings of appalling labour practices in *Labor and the Locavore: The Making of a Comprehensive Food Ethic*, published in 2014. Undocumented and guest workers on meagre incomes below the minimum wage and without overtime live in trailers on the farmers' property in constant fear of being deported. Farm bosses admit to preferring 'fresh-off-the-boat' immigrants with poor English skills because they find that as soon as they learn the language, they tend to get too uppity: "They get a little too Americanized. That's what I call it; they get Americanized, and then they get lazy," according to one farmer. A bill in the state senate in 2010 to expand farmworker protections was defeated by farmers who said that the bill would "decimate the 'personal touch' of farming."

Finally, the great devil of localists is the intensification of agriculture, but, as Desrochers and Shimizu argue, by concentrating the growing of crops in more suitable areas, permitting a greater output for the given area, and then trading and distributing them long distances, deforestation and soil erosion (due to trying to grow everything on the same unsuitable plot of land) are reduced. Desrochers and Shimizu at times lean toward too much of a faith in the price mechanism to deliver optimum environmental outcomes, but are also catholic enough elsewhere to even cheekily quote 19th Century Marxist economist Karl Kautsky from his seminal 1899 work on agriculture, *The Agrarian Question*: "As long as any rural economy is self-sufficient, it has to produce everything which it needs, irrespective of whether the soil is suitable or not. Grain has to be cultivated on infertile, stony and steeply sloping ground as well as on rich soils":

> Turning our back on the global food supply chain, and, in the process, reducing the quality of food produced in the most suitable locations will inevitably result in larger amounts of inferior land being put under cultivation, the outcome of which can only be less output and greater environmental damage.

Intensification of agriculture—production of more food in less area—has been a boon for reforestation in European countries since the early 19th Century, they note. Similar processes have occurred in all other developed countries, albeit of varying scale, alongside the abandonment of wood as fuel source.

Interestingly, spatial scientist Jon Fisher recently explored[71] how the latest global data[72] from the United Nations Food and Agriculture Organization shows that it "is simply not correct" that agriculture is consuming ever more habitat. Over the past 15 years, the world has *decreased* the total amount of land used to produce food, even as the global food supply has increased. As a product of this intensification process, significant tracts of agricultural land have returned to forest, more than the reverse process, which of course is a rare bit of good news as far as climate change is concerned. He is very careful to say that it is not all good news. This is the *global* state of affairs. There are many regions where the opposite process is happening, and not all agricultural intensification is sustainable. "Overall," he says, "agriculture has a long way to go to become truly sustainable."

But the important point is that dealing with the many complex problems within agriculture—from erosion to carbon emissions to marine dead zones, depletion of aquifers, and reduction of biodiversity—will require a combination of many complex solutions under the rubric of sustainable intensification. By contrast, oversimplified cross-the-board heuristics like 'buy local' can be actively harmful.

Beyond agriculture, local small businesses of any description frequently cannot afford to meet stringent environmental protection measures (which is why some large businesses can sometimes be quite happy with such legislation, or even lobby on its behalf, if it has the effect of blocking their smaller competitors). As Greg Sharzer, a left-wing critic of the new localism, notes in his book *No Local*, a UK government survey found that "whether or not a business had an environmental policy depended critically on size. Smaller businesses were least likely to ... safeguard the environment."

Sharzer also notes how closely contemporary localism echoes the perfect fair marketplace prescribed by the father of anarchism, Jean-Pierre Proudhon, wherein small-scale property is retained but large-scale property suppressed. This Proudhonism for the 21st Century defends small business over big business, believing the latter to be more ethical and eco-friendly, ignoring the market forces that affect both and that require the worker subjugation we are supposed to be opposing in the first place. Scale rather than capitalism becomes the enemy. It is a petty bourgeois criticism of size rather than a progressive, structural critique of the current organisation of our political economy. "Hidden beneath localism's DIY attitude is a deep pessimism: it assumes we can't make large-scale, collective social change," worries Sharzer.

Yet small businesses are not exempted from the unremitting drive to outcompete rivals. Indeed, again due to economies of scale, large businesses can sometimes be more able to pay higher wages and offer more substantial benefits than mom-and-pop operations. Sharzer notes how US companies with more than 500 employees pay a third higher wages than smaller firms, and are more likely to provide healthcare coverage. The most vitriolic opposition to efforts to boost the minimum wage in Canada for instance always comes from the Canadian Federation of Independent Businesses, the small-business lobby group.

Localism is ultimately presenting the instant gratification and easy option of ethical consumerism as a solution ("Saving the planet is as simple as buying artisanal camembert!") rather than the hard, years-long slog of society-wide organisation for structural change. It is also a very expensive solution for many, many working people, students and the elderly, involving a great deal of time and effort to educate oneself on different products. Are we really arguing that 'climate justice' will be achieved by demanding that single parents spend hours online researching this year's precise timing of the New Hampshire asparagus season?[73]

There are of course many products for which it really does make climate sense to be consuming the local version of a product—those places where the local production process is the most efficient. Quebec maple syrup, Florida oranges, Moroccan olives, and so on. Thus, assessment as to whether local is beneficial to the environment needs to be made on a case-by-case basis, or, as Desrochers and Shimizu put it: "Local when sensible." The globalisation of food (or production of any commodity) is not the enemy per se, although it could and should certainly be more rationally planned.

All this said, despite the blithe romanticism, wrongheadedness and, frankly, gullibility involved with localism, we needn't be too snarky about people engaged in local projects. "If you want to create healthy food for yourself or trade crafts, that's great. Making something yourself, whether it's a painting, a bicycle or a carrot, is a way to feel you've left a mark in a world where everything's bought and sold. If growing your own vegetables makes you feel better and helps you meet your neighbours, then you should do it," argues Sharzer. Indeed, all campaign work has to start somewhere. "Participating in a local DIY project can provide the strength and tools for community activism."

But if the goal is to stop ecological degradation, and climate change in particular, then the stakes are higher.

*

Drudgery For All!

In another example of the social backwardness of localism, these small-is-beautiful partisans regularly stress how de-intensification of agriculture can deliver more jobs and lower unemployment.

Naomi Klein, writing in *The Nation*[74], again offers an exemplar: "Another bonus: this type of farming is much more labour intensive than industrial agriculture, which means that farming can once again be a substantial source of employment."

Of course intensification of production by making employees work harder and faster for the same pay is obviously not something that progressives would celebrate. This is also an intensification of the rate of exploitation, an extraction of additional surplus value.

But why would the performance of dull, repetitive tasks by machines be something that we oppose? When new technology makes workers redundant, it is true, they are laid off. And if we wanted to, I suppose we could increase the number of jobs by getting humans to do once again what machines do. We could go back to hand-weaving cloth, milling grain with a mortar and pestle, and printing books by hand-cranked presses. We would see an enormous growth in employment if we went back to the rickshaw. (Urban theorist Mike Davis actually noted in his magisterial 2006 text on urbanisation, *Planet of Slums*, how while revolutionaries in the early 20th Century denounced the rickshaw, pulled by a "human animal," as the ultimate emblem of the degradation of labour and promised its elimination "come the glorious day," it remains a great employer in much of Asia, with rickshaw pullers numbering 3 million across the continent. The rickshaw sector is the second-largest provider of employment in Dhaka.)

Yet beyond a lamentably vast increase in drudgery, this ignores the real reasons why unemployment is a problem. In a just world, the reduction in the amount of human labour required for any task due to technological advance would be a reduction shared equally amongst us all, giving every one of us ever greater free time. The replacement of humans by automatic check-outs at the supermarket is a tragedy for the now unwanted workers who are thrown on the scrap heap, but in an egalitarian world, would we really want more people performing this boring task if can be performed by a machine instead? One of the great arguments for socialism has always been that the benefits from increased productivity would accrue to everyone, instead of being stored as profits enjoyed solely by the owners of businesses.

Again, this shows how conservative, how lacking in ambition is Klein's argument in favour of the de-intensification of labour. Despite her ostensible call for a focus on capitalism in her new book, there is an underlying, unconscious pessimism that actually doesn't really think capitalism can ever be done away with. The type of an egalitarian world of which I just spoke is utterly unrealistic, she and her ilk assume, so in place of socialism, we can only increase employment by a deintensification of production.

While socialists used to proselytise with slogans declaring that the better world to come will deliver ever more leisure time, the partisans of degrowth proselytise with slogans celebrating the pleasure of greater toil.

Certainly, great contentment and satisfaction can come from occasionally gardening, knitting a scarf, or binding a book by hand. And farmers and peasants may well take great pride in what they do. But at the same time, only one who has never worked as an agricultural labourer could imagine the back-breaking, seven-days-a-week tasks involved in farming to be anything other than a tribulation to be endured. No wonder those migrants who make the journey from the countryside to the city, from rural Saskatchewan to the bright lights of Toronto, or for that matter rural China or the DRC, look askance at the back-to-the-land zealots of contemporary urban bohemia and the dilettante publishers of that *McSweeney's* for hipster agricultur-alists, *Modern Farmer* magazine, with its coffee-table-ready analogue-camera-style photographs of heirloom tomatoes, backyard chickens and apartment-balcony beekeepers printed on 10-point-thick, matte-finished paper.

*

To be clear: there is of course nothing wrong with deer-hunting or home-pickling beetroot. And knowing your knots, orien-teering and how to build a fire from scratch are useful skills that can be employed on many occasions. (Blacksmithing and scything probably less so). I can still do the eight knot, bowline and round-turn-and-two-half-hitches that I learnt as a cub scout, which come in handy when tying down couches to the back of trucks. But let's not labour under the illusion that such practical activity, as enjoyable and rewarding as it can be, in any way represents a challenge to the status quo. (Even backyard chicken hatcheries are cool I suppose if that's your bag, although I did get a giggle recently coming across t-shirts that read "Goldman Sachs doesn't care if you raise chickens" [75]).

And the current obsession with collapse, with apocalypse from the Preppers, Peakists, and Transition Townies to McKibben, Jensen and Kingsnorth and the wider green left, is nothing less than a politics of despair in a period when genuinely transformative visions of an egalitarian economy appear foreclosed, a desire for a shortcut to a better world, a *deus ex machina* that can deliver the just society now that we 'know' that it is impossible for humans to organise collectively to do so. American left-wing economic commentator Doug Henwood, writing in *Catastrophism*, a 2012 collection of essays on the catastrophist zeitgeist, describes this mood best:

> We've not merely lost faith in transformative political projects, we often view them with fear: today's revolutionary will be in charge of tomorrow's firing squad, so let's junk ambition. Instead ... we project our transformative fantasies onto nature, but in a perversely destructive way. Nature will punish us for our ambition – 'industrial society,' the longing for material abundance, the urge to move beyond the small and local – by forcing us back to some hunter-gatherer purity, whether we like it or not. [76]

Sasha Lilley, writing in the same book, nails perfectly why this pessimism has arisen: "The appeal of catastrophism tends to be greatest during periods of weakness, defeat or organizational disarray of the radical left, when catastrophe is seen as the midwife of radical renewal."

Or, most succinctly, as American Marxist literary critic Fredric Jameson puts it in an epigram that has since been thrown up by philosophically-minded graffiti artists on the walls of cement underpasses the Anglophone world over: "It is easier to imagine the end of the world than to imagine the end of capitalism."[77]

Slovenian Marxist writer Slavoj Žižek repeats this Jameson quote so often that he is regularly confused as its author, a state of affairs 'the Elvis of Cultural Theory' is probably not unhappy about. He extends the argument in an interview in the 2005 documentary *Žižek!*:

> Think about the strangeness of today's situation. Thirty, forty years ago, we were still debating about what the future will be: communist, fascist, capitalist, whatever. Today, nobody even debates these issues. We all silently accept global capitalism is here to stay. On the other hand, we are obsessed with cosmic catastrophes: the whole life on earth disintegrating, because of some virus, because of an asteroid hitting the earth, and so on. So the paradox is, that it's much easier to imagine the end of all life on earth than a much more modest radical change in capitalism.

The primitivist novelist and top scything expert Paul Kingsnorth is perhaps the exemplar *ne plus ultra* of such defeatism, writing in *Orion*, the popular environmentalist magazine:

> If you think you can magic us out of the progress trap with new ideas or new technologies, you are wasting your time. If you think that the usual "campaigning" behaviour is going to work today where it didn't work yesterday, you will be wasting your time. If you think the machine can be reformed, tamed, or defanged, you will be wasting your time. If you draw up a great big plan for a better world based on science and rational argument, you will be wasting your time.[78]

UK philosopher Mark Fisher describes this pessimism, this acceptance of Thatcher's dictum of TINA—'there is no alternative'—as *capitalist realism*: "the widespread sense that not only is capitalism the only viable political and economic system, but also that it is now impossible even to imagine a coherent alternative to it." The leftist Fisher laments this as a "cancelling of the future" while the liberal Francis Fukushima famously celebrates it as the end of history. Either way, collapse porn is manifestly a product of belief in neoliberal foreclosure of alternatives.

We can also see this in the call to 'post-futurism' by Italian autonomist philosopher Franco 'Bifo' Berardi, in his most recent book, *After the Future*. In it, Bifo argues that the very idea of the future had its moment—*the Century that Trusted in the Future*—and that moment is over; a 'futurised' capitalism or socialism is *over*.

> In the second part of the 19th century, and in the first part of the 20th, the myth of the future reached its peak, becoming something more than an implicit belief: a true faith, based on the concept of "progress", the ideological translation of the reality of economic growth. Political action was reframed in the light of this faith in a progressive future. Liberalism and social democracy, nationalism and communism, and anarchism itself, all the different families of modern political theory share a common certainty: notwithstanding the darkness of the present, the future will be bright.[79]

But now the future has been cancelled. Yet, as pointed out by his anglophone editors, Gary Gonesko and Nicholas Thoburn, there is no escaping this end-point: "The point is not to revive the future in a new vanguard. The future was itself a highly suspect temporal form."

Instead, there is nothing to do other than, precisely as Kingsnorth prescribes, resign, withdraw, retreat:

> [A]ll along the modern times the myth of the future has been connected to the myth of energy; think about Faust, for

instance. This idea that the future is *energy*: more and more and more. More speed, more strength, more consumption, more things, more *violence* ... Everything has to be sacrificed to the growth—this abstract growth—of money, of value, *of nothing*. So, how can we withdrawal from this kind of craziness. I think that we have to act, and to live, in a post-futurist way which means we have to choose a slowness of pleasure—like the birds in the sky, like the flowers in the fields, they don't need to work, they don't need to accumulate, they don't need to possess. They need to have pleasure, to live, to live in time.[80]

Berardi goes on to call for an abandonment of all hopeful ideologies of the left, to be replaced with a "radical passivity":

I see a new way of thinking subjectivity: a reversal of the energetic subjectivation that animates the revolutionary theories of the 20th century, and the opening of an implosive theory of subversion, based on depression and exhaustion ... But exhaustion could also become the beginning of a slow movement towards a "wu wei" civilization, based on the withdrawal, and frugal expectations of life and consumption. Radicalism could abandon the mode of activism, and adopt the mode of passivity. A radical passivity would definitely threaten the ethos of relentless productivity that neoliberal politics has imposed.[81]

To my mind, far from being a project of resistance to capitalism, this is an acquiescence in the face of its perceived invulnerability. Running off to the woods and writing poetry instead of, say, unionising your workplace. Rather exactly the sort of deference and submission that the masters of the universe desire. Marxist cultural critic Alberto Toscano goes so far as to accuse these miserabilists of oblivious collaboration:

The anti-Prometheanism of the left, instead, is most often marked by melancholy or illusion. Melancholy: the sense that emancipation is an object better mourned than desired; that the price of our principles is prohibitive. Illusion: the persuasion that the powerless can prevail over the powerful without concentrating and organising their forces; the belief that the systems and capacities which now embody the dead labours of generations, and bear the traces of barbarisms past, can simply be abandoned or destroyed, rather than, at least in part, appropriated. Such attitudes channel, more or less unwittingly, that crucial counter-revolutionary tenet, according to which political violence and catastrophe is a consequence of imposing abstract ideas (liberty, equality, fraternity...) upon complex and refractory human material.[82]

While Klein, Kingsnorth, Berardi and their co-thinkers believe themselves to be critiquing capitalism, their modest proposals and catastrophist thinking—and in the case of the latter two, their strategy of retreat from struggle—actually fit within and contribute to a broader mood of abandonment of the possibility of any type of post-capitalist society. They are *captives of capitalist realism*. Despite their glee at civilisational *untergang*, the doomsayers inhere this neoliberal melancholia. Ostensibly and outwardly oppositionist, in reality, they are unconsciously submitting to what Derrick Jensen refers to as the *dominant paradigm*. They are 'anti-capitalists' who for all their bluster (and in the case of Jensen and his followers explicit calls to guerrilla-style violence) do not actually think or speak in revolutionary terms.

In a critique of James Cameron's 3D deep-ecology space opera blockbuster *Avatar*, in which the hero, paraplegic marine Jake Sully, is able to defeat interstellar imperialism only via Jensen-style primitivist rejection of his modernity, Fisher argues that the film is yet another example of 'corporate anti-capitalism' or

'Hollywood anti-capitalism', a *gestural* rather than substantive anti-capitalism where very frequently the villain in a movie is an 'evil corporation', where immorality (greed) rather than material, structural forces drive injustice.

No primitivist cliche is left untouched in Cameron's depiction of the Na'vi people and their world, Pandora. These elegant blue-skinned noble savages are at one with their beautiful world; they ... recognise that a vital flow pervades every-thing; they respect natural balance; they are adept hunters, but, after they kill their prey they thank its "brother spirit"; the trees whisper with the voices of their revered ancestors ... "There's nothing we have that they want," concludes Sam Worthington's Jake Sully of the Na'vi. Yet the Na'vi predictably seduce Sully, who quickly "forgets everything" about his former life on earth (about which we learn almost nothing, beyond the fact that he is a marine who got injured in the course of battle) and embraces the wholeness of the Na'vi way of life. Sully attains wholeness through his avatar Na'vi body in a double sense: first, because the avatar is able-bodied, and, secondly, because the Na'vi are intrinsically more "whole" than the (self-)destructive humans. Sully, the marine who is "really" a tree-hugging primitive, is a paradigm of that late capitalist subjectivity which disavows its modernity. There's something wonderfully ironic about the fact that Sully's – and our – identification with the Na'vi depends upon the very advanced technology that the Na'vi's way of life makes impossible.[83]

He laments how in *Avatar*, as in *Wall-E* and other contemporary Hollywood anti-capitalist films, what is available to us has been delimited to two wretched options precisely as a result of capitalist realism: "What is foreclosed in the opposition between a predatory technologised capitalism and a primitive organicism, evidently, is the possibility of *a modern, technologised anti-capitalism*. It is in presenting this pseudo-opposition that *Avatar* functions as an ideological symptom" [emphasis added].

Fisher concludes his review of Cameron's film by quoting a passage from Žižek's 2009 book, *First As Tragedy, Then As Farce*, which called on the left to abandon its quarter-century strategy of liberal moralistic blackmail, of seeing capitalism in precisely these moral terms. In this section of the book, Žižek criticises a 2008 public letter from the left-wing Bolivian president, Evo Morales, "Climate Change: Save the Planet from Capitalism":

> Today, our Mother Earth is ill ... Everything began with the industrial revolution in 1750, which gave birth to the capitalist system. In two and a half centuries, the so called "developed" countries have consumed a large part of the fossil fuels created over five million centuries ... Competition and the thirst for profit without limits of the capitalist system are destroying the planet. Under Capitalism we are not human beings but consumers. Under Capitalism Mother Earth does not exist, instead there are raw materials. Capitalism is the source of the asymmetries and imbalances in the world.

Žižek replies (with appropriately genuflecting caveats recognising that despite Morales' limitations, he remains at the very cutting edge of contemporary progressive struggle) that if there is one good thing about capitalism, it is precisely that Mother Earth no longer exists: "Fidelity to the communist[84] Idea means that, to repeat Arthur Rimbaud — *il faut être absolument moderne* — we should remain absolutely modern and reject the all too glib generalization whereby the critique of capitalism morphs into the critique of 'modern instrumental reason' or 'modern techno-logical civilization'."

We have been lied to by the likes of Naomi Klein, Bill McKibben, Derrick Jensen, Paul Kingsnorth and a few more whom I'll be getting around to giving a gentle poke in the following pages: Green anti-modernity does not threaten neoliberalism; *it is a manifestation of it!*

7

Escape from the Innovation Desert

There is one sense in which Bifo Berardi was correct about there being 'No Future' any more, or rather, recast by Mark Fisher, that there has been a "slow cancellation of the future": there does indeed seem to have been something of a pause to history for a couple of decades now, or at least a deceleration from the rapid rate of change experienced over the last two centuries. Fisher here is writing about popular culture, music specifically, and how the sense of future shock has disappeared, but he is also talking about the same disquieted ennui with respect to wider social forces:

> While 20th Century experimental culture was seized by a recombinatorial delirium, which made it feel as if newness was infinitely available, the 21st Century is oppressed by a crushing sense of finitude and exhaustion. It doesn't feel like the future. Or, alternatively, it doesn't feel as if the 21st Century has started yet. We remain trapped in the 20th Century.

This loosely expressed feeling that Fisher can't quite put his finger on is echoed further afield in the culture. The name of the somber Scottish indie band, We Were Promised Jetpacks, is not so much a petulant whine as a registration of a feeling of injustice over a contract not being honoured, and there's a bittersweetness to the retro-future nostalgia of the similarly titled book *Where's My Jetpack?: A Guide to the Future that Never Arrived* by roboticist Robert Wilson:

> It's the twenty-first century and let's be honest—things are a

little disappointing. Despite every World's Fair prediction, every futuristic ride at Disneyland, and the advertisements on the last page of every comic book, we are not living the future we were promised ... Where are the ray guns, the flying cars, and the hoverboards that we expected? What happened to our promised moon colonies? Our servant robots?

We are regularly told how people can barely keep up with the pace of change, but in fact it's Fisher and Wilson who have it right if we consider the question empirically. In 2012, US economist Robert Gordon produced a provocative paper suggesting that the last 250-odd years of strong growth and rapid technological advance may have puttered to a halt, that in fact for some three or four decades now, the west has been trapped in an unusually prolonged period of weak rates of investment in new technology and equipment, where the animal spirits of capitalism no longer seem to be pushing the productive forces forward. We forget that the enabling technology of the internet was actually developed in the late 60s to early 70s, and the case is the same for the mobile phone, and even genetic recombination was a Disco-era breakthrough. There has been very little that's genuinely new since, as opposed to the much more banal repeated iterations of gadget improvement. Economist Tyler Cowen makes a similar argument about this generation-long innovation desert in his influential 2011 essay "The Great Stagnation." Cowen argues that there's a simple reason for this: we've picked all the low-hanging technological fruit. That is, everything that we've invented up to now was relatively easy. What comes next in terms of innovation, whatever that is, is just a lot more difficult to come up with, and will more likely come about in incremental advances. A similarly widely shared essay in 2014, "The Golden Quarter," by British science journalist Michael Hanlon, again looked at how some of our greatest technological accomplishments occurred between

1945 and 1971. And such arguments are hardly the preserve of contrarian science writers or maverick economists. Governments long ago noticed the slow-down. The European Commission in particular is fond of setting up expert groups, "eliciting the views of major stakeholders," and commissioning top-flight consultancy firms to produce studies, surveys and trend analyses to explain why there has been a dearth of innovation and what policies can be implemented to turn this sorry state of affairs around. One of the EU's flagship initiatives is its 'Innovation Union', which it hopes will develop alongside monetary union, 'fiscal union', 'energy union', 'transport union' and other EU integration endeavours, although nobody really seems to know what an innovation union is.

What is interesting here is that for the right and centre, this mysterious slow-down in innovation is a source of great concern. And so the solutions presented usually revolve around tax incentives or subsidies, tweaking the patent system and strengthening intellectual property protection, deregulation, or yet another excuse to 'liberalise the labour market'. Meanwhile, the left seems hardly to have noticed this at all. The innovation desert never features alongside increasing inequality, anti-war organising, the science of climate change or Palestine solidarity and so on that typically feature as the subjects of public talks, seminars or conferences where left-wing activists, trade unionists, and academics gather. And yet we should have so much to say on this front. In her superb 2013 book, *The Entrepreneurial State*, American neo-Keynesian economist Mariana Mazzucato carefully and empirically shows how the private sector is in fact a timid, conservative economic actor in comparison to the state. The public sector, she demonstrates, does not just provide the bulk of research, development and commercialisation funding, but has always been the leading force behind the 'mission-oriented' introduction of the most radically innovative new technologies, from the mass production system, through aviation and computation, to biotechnology, satellites and space exploration, and now low-carbon energy and dematerialisation. Microchips, the internet, GPS, mobile communication, even touchscreen technology and Siri? All government funded. Almost 75 percent of all genuinely innovative new drugs, as opposed to 'me too' drugs (medicines that are chemically almost identical to existing drugs, with only minor differences)? That's the US National Institutes of Health you have to thank.

Moreover, that 'Golden Quarter' of innovation, from 1945–71, is almost perfectly coincident with the postwar period of strong unions and strong social protections that the French call 'Les Trentes Glorieuses' (45–75), and the desert of innovation since then is coincident with the rise of neoliberalism. This combines with Mazzucato's analysis in a way that could permit a left that re-embraced modernism to deliver a pretty good solution to the problem of the innovation desert. But the crucial thing here is not just — or not even — the cuts in public spending on research and development, or the reduced proportion of spending on pure research compared to that on applied research, although these are subjects that the left should have better answers to than the right. The crucial change that occurred during the Golden Quarter was the decommodification that happened over the postwar period. That is, public-sector products, services, research, were not developed for their exchange value (for the most part), but for their utility to society, their use value. Today, even public services must embrace an entrepreneurial spirit, show a profit, show a healthy return on investment and so on. There has been a recommodification of what had been decommodified.

The Keynesian welfare state and public-sector driven investment decisions were far from perfect, sure. Paternalist even. And it is a mistake when many on the liberal-left, like former Clinton labour secretary Robert Reich and Nobel laureate economist Paul Krugman, call for a return to those days. (How could we do this anyway? Keynesianism ended not because Margaret Thatcher and Ronald Reagan were elected, but because Keynesianism was broken. Thatcher and Reagan were just symptoms, and their policies were largely just intensifications of what had been pioneered by Harold Wilson and Jimmy Carter). Neoliberalism is not countered by uncritically embracing the state. Nonetheless, the dirigisme of this period had undeniable benefits. Public healthcare, antibiotics, vaccines; everyone gets an indoor toilet, a washing machine, a fridge and a TV; we put a man on the fucking moon. It was without doubt the high watermark of civilisation.

That's the public sector side of things. But neoliberalism has also seen a steady financialisation of the economy as well. GM and Ford don't make cars anymore; they're financial institutions. This means it is a safer bet to make money from financial innovation than from actually taking the risk of building stuff. All this means that *both* sides of the economy—public and private sectors—are structurally inhibited from engaging in radical innovation as a result of neoliberalism.

Once upon a time, the left promised that socialism could provide a better future than capitalism, not merely due to the elimination of inequality and crisis, but also because a planned economy could materially outperform an unplanned one. The set of products and services that are profitable is much smaller than the set of products and services that are useful to humanity. That is, we promised more innovation, faster progress, *greater abundance*. One of the reasons I believe that the historically fringe ideology of libertarianism is today so surprisingly popular in Silicon Valley and with tech-savvy young people more broadly; with billionaire tech entrepreneurs like Elon Musk, Peter Thiel and Jimmy Wales; with hackers, whistleblowers and civil liberties campaigners like Julian Assange, Edward Snowden, Chelsea Manning and Glenn Greenwald; and with members of digital rights NGOs like the Electronic Frontier Foundation and Europe's

La Quadrature du Net, is that libertarianism is the only extant ideology that so substantially promises a significantly materially better future.

We see an interesting parallel to the unrealised promises of a high-tech future with the erroneous early and mid-20th Century predictions of an ever-shrinking working week. In a short essay, "Economic Possibilities for Our Grandchildren," written in 1931, the second year of the Great Depression, Keynes confidently predicted that by 2030, technological advance would "solve mankind's economic problem." Progress would enable us to work just three-hour shifts a day or 15 hours a week while enjoying a standard of living richer than the very richest knew in his time:

> I would predict that the standard of life in progressive countries one hundred years hence will be between four and eight times as high as it is today. There would be nothing surprising in this even in the light of our present knowledge. It would not be foolish to contemplate the possibility of a far greater progress still.

In 1974, the great science fiction author Isaac Asimov wrote an essay for the *New York Times* trying to imagine what the future 50 years on would look like. From video telephony to driverless cars, many of his predictions have come uncannily true. But one prediction is glaringly wrong:

> The world of A.D. 2014 will have few routine jobs that cannot be done better by some machine than by any human being ... [M]ankind will suffer badly from the disease of boredom, a disease spreading more widely each year and growing in intensity. This will have serious mental, emotional and socio-logical consequences, and I dare say that psychiatry will be far and away the most important medical specialty in 2014.

The lucky few who can be involved in creative work of any sort will be the true elite of mankind, for they alone will do more than serve a machine. Indeed, the most somber speculation I can make about A.D. 2014 is that in a society of enforced leisure, the most glorious single word in the vocabulary will have become *work!*

How do we return to an age of steady innovative advance and ever increasing leisure time? In essence, we have to repeat the trick of how we achieved the Golden Quarter/*Les Trentes Glorieuses* in the first place.

And how did we do that? To answer such a question, we'll need to take a brief economic history detour.

*

Let's remember that the first genuine 'Keynesian turn'—in form if not by name—initially occurred not after World War Two, but under Mussolini and Hitler. Between the manifest cataclysm of laissez-faire capitalism and the uncertain promise of Communism, it was the corporatist innovations of fascist economics that prefigured the Scandinavian, Swiss and American demand-management economic policies a few years later which were then embraced across western Europe after the war by governments of both left and right.[85]

Both fascist governments were able to overcome economic crisis by stimulating aggregate demand through massive investment directed by state agencies and the establishment of extensive and generous social programmes. Italy saw the establishment of work creation programmes involving house construction, marsh draining and highway, canal, railroad infrastructure roll-out. Government assumed control of credit allocation, with the fascist regime's *Istituto per la Ricostruzione Industriale* in this period ultimately controlling 77% of pig iron production, 45% of steel production, 80% of naval construction and 90% of shipping.

When private capital was frightened of investing, the state became the investor of last resort. The regime also oversaw banking reform and nationalised the Bank of Italy. It also introduced the 40-hour work week, health insurance, paid national holidays, disability and old-age pensions, maternity benefits, and leisure-activity subsidies. State expenditure doubled between 1922 and 1933.

Meanwhile in Germany, when we ask ourselves how it is that so many people could tolerate the totalitarian barbarism of the Nazis, we have to understand that for many ordinary Germans who were not from minority groups or the left, the daily experience was not one of terror, but, as Gotz Aly argued in a 2005 essay in Der Spiegel, a 'warm and fuzzy' or 'feel-good' dictatorship: *'Die Wohlfuehl-Diktator'*[86]. Employment returned through a similar stimulation of demand, while the regime delivered family and child supports, decent pensions, free access to higher education and even cheap tickets for the theatre and concerts. There was an economic leveling and even social mobility. The well-being of the national community, *'volksgemein-schaft'*, was all-important. Life was indeed better.

But this all this was made possible not just by high rates of corporate taxation, near-autarkic capital controls and the state assumption of investment decision-making, but also ultimately slave labour (by Jews, eastern Europeans, political dissidents, homosexuals and POWs—amounting to 20 percent of the German workforce at its peak) and the conquering and economic draining of other economies.

Meanwhile, fascism also promised and delivered an end to the widespread class conflict of earlier decades. Independent working-class institutions were obliterated. Business leaders embraced what had previously been seen as fringe parties due to the social peace they imposed. Fascism is at base a response to crisis and strife.

Scandinavia in this period had also been the site of recurrent labour unrest. Labour relations in Norway and Sweden were home to strike rates amongst the highest in the world, and frequent violent clashes. The worst of which, a paper-mill strike in Adelen in 1931, was suppressed by the army. In 1938, fearing further violence and radicalisation, and the socialism that was around the corner, a historic compromise between labour and capital was reached at Saltsjöbaden. For similar reasons the *Arbeitsfrieden* (labour peace) agreement was achieved in Switzerland in 1937.

A similar social and economic reorganisation, with local particularities, to what had been pioneered in the fascist countries was adopted. In Sweden, as the late historian Tony Judt pointed out in his magisterial history of Europe since World War Two, *Postwar*, Swedish social democracy's concept of the 'people's home' — *Folkhemmet* — was appropriated from the nationalist political theorist Rudolf Kjellen, who in turn was inspired by the social-reformism of Otto von Bismarck's conservative Realpolitik. Via Kjellen's concepts of folk not *klass*, as Judt argues, meant Sweden's *Sveriges socialdemokratiska arbetareparti* was able to undercut the right's communitarian appeal and neutralise support for fascism.

Sweden certainly benefited from an economy that had not been devastated by war as other western European countries had been in World War One, leaving it more room to manoeuvre, but she also benefited directly from the upturn in the German and American economies, which had also embraced a demand-management approach.

Under the New Deal, the US Treasury embraced deficit public spending while labour and capital reached a 'Grand Truce' akin to Saltsjöbaden in Sweden and the *Arbeitsfrieden* in Switzerland.

Although some of the bright young things recruited to the US Treasury may have been Keynesian true believers, Roosevelt himself was no idealist. The president had concluded that governments had no option but to turn away from laissez-faire liberalism for a brief period in order to stave off revolution.

Canadian economists Leo Panitch and Sam Gindin in their recent history of the construction of the international postwar economy, *The Making of Global Capitalism*, uncovered a remarkable quote from the US president, writing in a letter to a friend that there was "no question in my mind that it is time for the country to become fairly radical for at least one generation. History shows that where this occurs occasionally, nations are saved from revolutions." The authors note that he admired the ultimately doomed municipal social democracy of the experiment of Red Vienna in the 1920s and its public-housing projects for having "probably done more to prevent Communism and rioting and revolution than anything in the last five years."

And indeed, as the war ended, in many parts of the continent ordinary people were armed, with millions expecting that the world that would be built after the conflict would be radically different and thus taking things into their own hands.

*

In the absence of government, and following desertion by factory and land owners, local committees were set up to administer workplaces and neighbourhoods, organise food supplies and maintain order—often a rough justice toward collaborators. In 1944, a popular insurrection in Ragusa, Sicily, was put down by the army, and in 1948, a three-day general strike crippled the country.

While arriving Allied armies worked to disband these councils and committees, US forces cabled home to report their deep concern at the disturbances that they feared could quickly evolve into revolutionary situations. Their own forces were not immune to the rebellious sentiment. In 1946, a thousand US GIs in Paris protested down the Champs-Élysées and a New York paper the same year fretted in a report from Nuremberg that American soldiers had been infected with "strike fever." Across the continent, Communist Party memberships soared. And of course civil war in Greece broke out between Communist partisans and nationalists, with Britain backing the fascists who not months before had been the enemy. Socialism once again was a greater fear than fascism.

It was clear that, as British Conservative MP Quentin Hogg told the House of Commons as early as 1943: "If you do not give the people reform, they are going to give you revolution."

And so we find that the post-war welfare state, full employment and concomitant high-wage truce with labour was constructed as much by parties of the right as by social-democratic governments.

The 1947 programme of Germany's Christian Democratic Union declared: "The new structure of the German economy must start from the realisation that the period of uncurtailed rule by private capitalism is over." It went on to argue that the "capitalist economic system" had not served the German people, and called for a "new order built right from the ground" based on "an economic system of collective ownership." And the 1944 manifesto of France's Christian democratic Popular Republican Movement backed what it described as a "revolution" to create a state "liberated from the power of those who possess wealth." It backed the nationalisation of industry and the banks and supported the programme of the *Conseil National de la Résistance*, which envisioned a social-democratic planned economy after liberation. In France, it was Gaullism rather than social democracy that became the dominant force, but Gaullism was just as committed to demand management as its left opponents. In Austria, the Socialists and the conservative People's Party reached an understanding that was intended to prevent the sharp class conflict of the decades before the war. This delivered child care, unemployment insurance, public pensions, child support, universal medical provision, public education and subsidised transport. While it had experienced a typical Western European welfare state for decades, it was not actually until 1970 that the country saw its first left-wing chancellor, Bruno Kreisky.

In the UK, as in Belgium and the Netherlands, it was indeed their Labour Parties that built the welfare state, with demand-manipulating fiscal policy, progressive taxation and some level of nationalisation, and with as much cooperation as opposition from the mainstream right. As James Meek writes in an extensive essay in the *London Review of Books* on the history of public housing in the UK, "the Tories started a race with Labour over who could build more houses."[87] Conservative Prime Minister Harold Macmillan at one point said: "Toryism is paternalist socialism" and advocated a "planned capitalism" with substantial public ownership.

Certainly what aided the process on the continent was that the old conservative political class leading Europe had been utterly discredited by its association with fascism. The new Christian Democratic parties were composed initially of anti-fascist Catholics, some of whom had been partisans alongside Communists and socialists, allied to Christian trade unions. With the old order razed to the ground, amid the rubble, a new order could be built, even if the traditional right would quietly come in from the cold to absorb these parties in the years to come.

And of course, it was that grand American villain of the right, Richard Nixon, who famously proclaimed: "We are all Keynesians now."

This is not to say that there were no differences between social democrats and Christian Democrats/conservatives. Compared to the cosy consensus between today's centre-left and centre-right on economic questions, even amidst the post-war Keynesian consensus, the contrast in party programmes and delivery was stark. Just to take the example of Austria, the first

'Red Chancellor' Kreisky oversaw the extension of employee benefits while cutting the working week to 40 hours, decriminalised abortion and homosexuality, legislated formal equality for women and introduced maternity leave. He also extended public ownership to the point where Austria had one of the largest nationalised sectors outside of Communism.

But the point is that Europe had to pass through decades of war, violent class-conflict, revolution and holocaust, for there to be cross-party consensus that mass unemployment, what they believed to be the catalyst of all this strife, was a treacherous menace that had to be avoided despite the not insubstantial cost of the welfare state and high wages.

This is where the welfare state came from: Not from enlightened social democrats on high, but from grubby conservative elites petrified that if they didn't do something pretty radical, they would lose everything.

It is a cliché to make reference in political writing to the words of the nephew of Don Fabrizio Corbera, Prince of Salina, from Giuseppe Tomasi di Lampedusa's 1958 novel about the Italian Risorgimento, *Il Gattopardo*. But the epigram is only really accurate as an allegory for the historic compromise of the capitalist class after the war: "Unless we ourselves take a hand now, they'll foist a republic on us. If we want things to stay as they are, things will have to change."

*

To return to the question of rapid innovation and expanding leisure, we can now see how the Golden Quarter/*Les Trentes Glorieuses* was a direct product of the militancy of the working classes of the previous thirty years, an epoch of revolution, war, depression and the barbarism of the Second World War. We won social reforms—and the accompanying processes of state-led decommodification that bootstraps innovation—because, as articulated perfectly by Franklin Roosevelt and Quentin Hogg, the elites were frightened that without them, they would face social revolution and lose everything. Thirty years later, the Keynesian consensus broke down because of the contradictions at its heart, and the working class as a self-aware class was broken utterly in the 80s and 90s. To win anything close to this scale of popular advance in the standard of living again, or surpass it, this historic defeat has to be reversed, which means that the same scale of mobilisation of a militant *class for itself* will be necessary once again.

Profit drives efficiency gains, certainly, but there is a reduced need to innovate if it is easier just to repress your workers more, to extract more surplus value. Over the course of the Great Stagnation, as labour's share of wealth has steadily declined, pressure to innovate has declined in parallel. That is to say: capitalism, at least in part, needs a militant working class to force it to innovate! Left to its own devices (i.e. without opposition), it stagnates.

And here we have a beautiful symmetry: not only has the working class lost its sense of being a militant class for itself since the 70s, but the bourgeoisie is adrift since then as well, decadent even, its bearings lost in the absence of its historic enemy. This is another reason why the working class is not just the liberator of itself, but of all mankind. It is the universal class. We are familiar with Marx's famous apophthegm: "The history of all hitherto existing societies is the history of class struggle." Well, inverting that you get that without class struggle, there is no history. History (innovation) has paused.

*

Such analysis should be tempered with the recognition that predictions about the future, as Asimov and Keynes show, are invariably, mightily, laughably wrong. To be sure, we still have a profusion of wonders yet to be discovered and invented within the extant fields of genetic engineering, nanotechnology, neuroscience (including brain-machine interfaces that already are allowing people to move prosthetic limbs with their minds, an absolutely remarkable achievement), high-throughput computational discovery of novel materials, personalised medicine, machine learning, the awesome and ominous predictive power of Big Data. There are many analysts who believe these and other fields (such as the more intractably frustrating fields of quantum computing and artificial intelligence research) will sooner or later lead us out of the innovation desert.

But even here, what will be produced in such fields, and others yet to emerge, will be so much more circumscribed *than what there could otherwise be*. And however imprecise my worries about the future turn out to be, it is evident that the neoliberal species of capitalism unnecessarily bridles innovation, as does capitalism more broadly; it should be remembered that the left has in its toolbox a series of implementable policy responses to the innovation desert; and, most importantly, it is clear that if the left is to experience any revival, it must return the question of innovation—or to use more traditional leftist language, the further development of the forces of production—to the heart of its analysis and its programme for change. We need to stop being dismissive of the perfectly reasonable *aspirational* outlook of the vast majority of people. Instead of sneering at people's desire for a better life, we can promise one so much better than the capitalists can.

The true revolutionary today is one who speaks of optimism, big, bold ideas, universal values and ambitious, globe-straddling, liberatory projects both technological and political. To stand in opposition to the inadvertent ecological calamity of capitalist unreason is to be, unapologetically, a heavens-storming modernist!

8

Frankenpolitics

There is perhaps no greater symbol of contemporary green-left progressives' retreat from reason, evidence and modernity than the global movement against genetic modification. The movement has its adherents on the right, certainly, but the movement is largely a phenomenon of the left, with its origins in environmental campaign groups in the UK and Canada in the mid-1990s. But is anti-GM campaigning actually that left-wing?[88]

In the spring of 2012, British activists from a pop-up campaign group, Take the Flour Back[89], announced that they were going to 'decontaminate'—or, rather, tear up—GM wheat being tested by the Rothamsted Research institute, one of the oldest agricultural research bodies in the world. The grain being tested gives off an odour that repels aphids, and also attract wasps that parasitise the insects. As a result, the wheat, developed by publicly funded scientists, would require less synthetic pesticide – a development that is hardly likely to deliver profits to the pesticide manufacturers.

The next month, a 30-year-old research project in Italy, involving transgenic olive trees, cherry trees and kiwifruit vines—one of the longest-running GM trials in Europe—was ordered destroyed with only a few days' notice by a court under pressure from an anti-GM group, the Genetic Rights Foundation.[90]

The hoped-for result of the non-profit research, led by plant scientist Eddo Rugini at the University of Tuscia, not concocted by any moustache-twiddling villains at Monsanto HQ, would again be a limit in the need for pesticides.

Students and colleagues stood by the aging Ruggini in solidarity amid the olive groves, but still the bulldozers arrived to rip up his life's work. The elderly scientist was devastated.[91] Colleagues encouraged him to move to the United States, where he'd received offers of work and where the mood is less fearful, but he replied despondently that he was simply too tired now.

When I was in Mexico in the summer of 2012 investigating a series of bombings of nanotechnology researchers by eco-anarchists[92], I met a husband-and-wife team of molecular biologists working at a public university whose lab had twice been the target of anti-GM arsonists of a similar ideology to the nanotech bombers. The scientists described themselves as socialists and strong supporters of the mass Yo Soy 132 protests that year against electoral corruption by the right. They were also keen to say how they were very much opponents of Monsanto and agribusiness. Indeed, they said how they were frustrated that historically a great deal of crop research had been performed by Northern experts with little knowledge of the needs of Mexican farmers and consumers. So their aim instead was to develop transgenic crops resistant to drought and insects that built on local knowledge. Their work developing GM crops was a product of their belief in social justice, not an exception to it.

Is it beyond the imagination of anti-GM activists that genetic modification could be used for public benefit instead of private profit? The activists may well be sincere in opposing social injustice, but all the same, they think that these problems arise from something inherent in the technology. In so doing, the complaint is in fact not the business practices of Monsanto, or even capitalism, but technology and progress itself.

The left used to be quite clear that technologies used in the context of colonisation and exploitation in another political and economic context could be liberatory. We didn't want to do away with industry, but rather capture it and run it democratically.

There is nothing intrinsically malign about any particular technology outside of the context in which it is used. Knives can be used to chop cauliflower or to murder Tutsis and Hutus.

Between the tool that we all use every day to cook with and the horror of Rwandan genocide, absolutely nothing in the technology of the knife has changed. All that has changed is the political economy.

*

Anti-GM activists are too often guilty of a variety of cherry-picking that *New York Times* environment correspondent Andy Revkin has called 'single-study syndrome': embracing a single study or handful of studies that fly in the face of the wider consensus.

In 2011, the *Journal of Coastal Research* published a study that purported to show that global sea-level rise has actually slowed since 1930. Subsequently debunked by climate scientists and research from the US Geological Survey, this one article has nevertheless been seized on by conservative climate sceptics[93] around the world, and the authors, James Houston, retired director of the US Army Corps of Engineers' research centre in Vicksburg, Mississippi, and Robert Dean, emeritus professor of coastal engineering at the University of Florida in Gainesville, have toured US climate-sceptic conferences.

While the US green left was correctly quick to condemn the Tea Party embrace of this single climate-sceptic paper, the anti-GM lobby continually refers to last year's 'study' by Gilles-Eric Seralini claiming to show how GM corn causes cancer in rats and infamously discredited for its jaw-droppingly poor method-ology. GM opponents say that Seralini has been the victim of an "orchestrated media campaign" to "silence" him, paid for by the biotech industry. But the criticism came from all quarters, not just Monsanto and friends.

In a rare joint statement, the French national academies of agriculture, medicine, pharmacy, sciences, technology and veterinary studies denounced the study as a "scientific non-event" that "spread fear among the public." The country's Higher Biotechnologies Council (HCB) declared: "The study provides no scientific information regarding the detection of any health risk," while the National Agency for Food Safety (ANSES) said simply but witheringly: "The data are insufficient to establish scientifically a causal link."

At last year's annual meeting of the American Association for the Advancement of Science (AAAS) — the 125,000-strong professional association of US scientists — President Nina Fedoroff said she was now "scared to death" by what she described as an anti-science movement. "We are sliding back into a dark era," she said. "And there seems little we can do about it."

She spoke about academics and government researchers being stalked and intimidated over their research into climate change; email hacking, Facebook campaigns calling for them to be fired; expensive PR efforts by oil companies and think-tanks working to discredit the concept of anthropogenic global warming; and toe-curlingly shameless displays of scientific illiteracy by prominent Republican politicians.

We are familiar with these sort of attacks on science from the right, of blimpish Tory climate denialism and Louisiana textbooks telling children that the existence of the Loch Ness Monster is proof that evolution is wrong. But Fedoroff was just as frightened of the vandalism, intimidation and violence directed towards biotechnology researchers from the green left. "I am profoundly depressed at just how difficult it has become merely to get a realistic conversation started on issues such as climate change or genetically modified organisms," she continued.

Have Monsanto and Syngenta managed to bribe the entire French and American scientific establishments? Well, if you read GMWatch, you probably think so. The leading anti-GM website actually believes the AAAS to be "captured from the top down." This is as absurd and poorly argued as right-wing accusations from denialist bloggers like Watts Up With That's Anthony Watts that the UN's Intergovernmental Panel on Climate Change has been captured by Greenpeace.

It should be a deep embarrassment to progressives, but the truth is that anti-GM activists are as guilty of anti-scientific thinking with regard to their pet subject as the Koch Brothers or the American Enterprise Institute are on global warming.

While there are a tiny number of scientists that question anthropogenic global warming, the overwhelming consensus is that human activity is responsible for the sharp increase in atmospheric CO_2 over the past two centuries. Equally, while Gilles-Eric Seralini may be a professor of molecular biology at the University of Caen, the overwhelming scientific consensus is that there is no risk to human health or the environment from GM as a suite of techniques. Pointing at Seralini's work and shouting "Look! Science-y!" ain't enough.

This 2013 statement from the AAAS on the subject really does give a sense of how anti-GM is as fringe as climate denialism:

The science is quite clear: crop improvement by the modern molecular techniques of biotechnology is safe. The World Health Organization, the American Medical Association, the U.S. National Academy of Sciences, the British Royal Society, and every other respected organization that has examined the evidence has come to the same conclusion: consuming foods containing ingredients derived from GM crops is no riskier than consuming the same foods containing ingredients from crop plants modified by conventional plant improvement techniques.

*

Nonetheless, opponents will regularly claim: "GM isn't a isn't a useful technology anyway," or "We don't need it."

Let me ask you: is the mass-production of insulin useful?

Bacteria were some of the first organisms to be genetically modified by researchers. One of the earliest such instances of this was the insertion of the human insulin gene into E. coli bacteria to produce synthetic human insulin, or 'Humulin'—indistinguishable from the pancreatic human version, developed by San Francisco biotech firm Genentech and first commercialised in 1982. Do go ask those with diabetes how useful this GM product is. Sadly, without any evidence other than the claim that humulin is 'unnatural', anti-GM groups like the US Organic Consumers' Association, Natural News, GM Watch and the Center for Food Safety want diabetes patients to opt for so-called natural animal insulin purified from the pancreas of cows and pigs over what they feel is the 'Frankenmedicine' variety, claiming that doctors have been coerced into "forcing patients off natural insulin" by Eli Lilly.

A range of human proteins helpful for a variety of medical conditions have also been produced since the 1980s through related processes, including blood-clotting factors for haemophiliacs and human growth hormone to combat dwarfism—proteins that were previously derived from cadavers and as such risked transmitting diseases. Hepatitis B and HPV vaccines have been developed using genetic engineering.

Or how about cancer modelling—is that useful? Would these critics of GM say that "we do not need" the OncoMouse, the laboratory mouse genetically modified to carry a gene that when activated increases the chance that the mouse will develop cancer, thus making it extremely useful for cancer research?

Looking to the near future, if via the development of self-destructive GM mosquitoes, we can do away with mosquito-borne diseases such as malaria, dengue fever, and chikungunya, amongst others, isn't that useful? This is not science-fiction dreaming. A 2013 trial deployment in Brazil of mosquitoes engineered by a small Oxford biotech company to be sterile showed an incredible 96 percent suppression of dengue mosquito, Aedes aegypti. The trials were organised by the University of Sao Paolo and funded by the government.[94]

Efforts to trial the GM mosquitoes in Florida last year however ran up against furious residents het up by fibs by environmental groups, despite the real public-health danger presented by the steady northward spread of mosquito-borne diseases as the climate changes.

Mosquitoes cause more human suffering than any other organism. Over a million people die every year from diseases spread by our ancient buzzing companion. If the sterile insect technique proves to be as successful as hoped, this will be one of the greatest advances in the history of our species, up there with the discovery of antibiotics and vaccines. The Left should be fighting to ensure that all parts of the world affected by mosquito-borne diseases get full access to GM mosquitoes, rather than just those regions that can afford the technique, and not campaigning to stop the trials.

And let's just ask farmers themselves whether they find GM useful. In 1996, when Argentina first approved the cultivation of GM crops, it refused to grant Monsanto a patent for its Roundup Ready soybean seeds. The country has loose intellectual property rights for plant varieties, an intellectual property regime that has been the source of longstanding battles between Argentina and the company, with the government at one point denouncing Monsanto's aggressive patent protection efforts as 'extortion'. As a result of the impasse, 'pirate' use of the product soared.

The country is now the third largest producer of GM food in the world after the US and Brazil. By 2005, while some 80% of the country's soybean acreage was planted with Monsanto's Roundup Ready, only 28–50% of soybeans were 'legally' sold.

Meanwhile in Pakistan, far from being a case where farmers were coerced into the use of GM, widespread smuggling of corn, wheat, cotton and vegetable seeds forced the government to give up on a completely ineffective ban on the technology. As of 2012, some 90% of the cotton grown in the country came from genetically modified seeds, while few farmers pay any royalties. Similar piracy occurs in Brazil. Why? Because of the generous savings accrued from the reduced inputs that are required.

*

Genetic modification is fundamentally dangerous, argue anti-GM protesters, because "we're meddling in things that we only half understand." Well, we knew even less about genetics when we started artificial selection (a.k.a. breeding) around 10,000 years ago.

Plant breeders from the beginnings of crop cultivation sought out desirable traits in wild plants, unpredictably shuffling the genes of species via cross-breeding. It's not true of course that cross-breeding is the same as modern genetic modification, but tangerines and nectarines for example are cultivars that certainly don't exist in nature, and broccoli was engineered from a relative of the cabbage by the ancient Etruscans. Cauliflowers are no more 'natural' than Flavr Savr tomatoes. But the difference with cauliflowers is that we are so used to them that we think of them as 'natural', and hence 'good'.

To say that we are developing organisms "that have never been seen in nature before" is true. But the poodle, achieved through selective breeding, had also "never been seen in nature before."

In the middle of the last century, our initial understanding of genetics allowed us to use chemicals and radiation to begin to accelerate the genetic changes we desired, resulting in products that were more nutrient-rich, hardier, and more drought-resistant. Then in the 1970s, modern molecular genetics and the invention of large-scale DNA sequencing permitted a profound improvement in our understanding of genetics. This in turn resulted in the creation of new methods that allow the very precise addition of useful traits to organisms.

The difference between ancient and modern genetic modification is, you could say, just the level of precision.

Another argument made by critics is that GM will not solve world hunger as some GM boosters claim. This at least is correct. There is – right now – more than enough food to feed the world's population and then some. It's not underproduction that's the problem, but lack of an egalitarian distribution. Only by radically transforming our economic system will we truly do away with hunger.

But between now and the glorious day when this is achieved, why can't we also, for example, improve nutrition through technology? Dietary micronutrient deficiencies—the lack of vitamin A, iodine, iron or zinc—produce marked increases in blindness, susceptibility to disease, and child mortality around the world.

Golden Rice, a variety of rice genetically engineered to be enriched with beta-carotene, a precursor of vitamin A, was created with the aim of improving the nutrient-density of meals in those areas of the world where rice is all people can afford. After its development by publicly funded researchers at the Swiss Federal Institute of Technology and the University of Freiburg, biotech firm Syngenta subsequently developed a variety that produced 23 times more beta-carotene than the original Golden Rice. It is to be offered to poor farmers royalty-free and farmers may keep the seeds for replanting.

Greenpeace opposes its release as it could open the door to wider deployment of GM technology, and the venerable anti-GM campaigner Vandana Shiva argues that in focussing on vitamin A deficiency, the promoters of Golden Rice will prevent the wider, necessary discussion about the causes of malnutrition.

Globally, some 10 million children under the age of five die every year—a large number of them from diseases that could be prevented by better nutrition. Arguing that they need to die so that people will wake up to the barbarities of capitalism is itself barbaric. And yet Shiva keeps being invited to lefty conference after lefty conference! The left needs to provide a platform to Vandana Shiva like it needs to provide a platform to the former head of the proto-fascist English Defence League, Tommy Robinson.

But what about the campaigners' favourite GM horror story, the infamous sharp rise in farmer suicides in India since the introduction in 2002 of varieties of cotton genetically modified to express Bacillus thuringiensis (Bt) genes to produce resistance to bollworms?

A disingenuous 2005 PBS Frontline documentary suggested that the use of GM seeds from Monsanto and Cargill have led to increased debt burdens, with farmers forced into indentured labour to pay off loan sharks. But in 2008, the International Food Policy Research Institute[95], an independent agricultural research institute that has been sharply critical of multinationals, mounted the most extensive investigation into the subject, sifting through peer-reviewed journal articles, official and unofficial reports, media reports and broadcasts, and found that "there is no evidence in available data of a 'resurgence' of farmer suicides" since 2002, and sharply criticised "media hype ... fuelled by civil society organisations."

The study found that the phenomenon of farmer suicides has been largely constant since 1997, arguing that the reasons for the growth in suicides—which is occurring across society—is complex, involving indebtedness, poor agricultural income, a downturn in the economy that had caused the re-ruralisation of urban-dwellers, the absence of counselling services, inadequate irrigation and the difficulty of farming in semi-arid regions. The decision by the government to reduce minimum support prices, World Trade Organisation policies and continued western cotton subsidies that make local cotton uncompetitive must also be taken into account.

A parallel investigation from economist K Nagaraj of the Madras Institute of Development Studies noted that a "mono-causal explanation of this complex phenomenon would be totally inadequate." The author argues that suicides are concentrated in regions with high and predatory commercialisation of agriculture and very high levels of peasant debt. He notes that cash crop farmers are more susceptible than food crop growers and argues that one must look to a massive decline in investment in agriculture, the withdrawal of bank credit at a time of climbing input prices, a crash in farm incomes, growing water stress and efforts toward water privatisation. Remove Bt cotton from the equation and all these other factors remain untouched. Focussing on genetic modification and ignoring the real causes — which are, as Nagaraj puts it, an "acute agrarian crisis in the country – and the state policies underlying this crisis" — is a dangerous distraction.

Meanwhile, a 20-year study at 36 sites in six provinces in northern China published in the journal *Nature* last January suggested that the deployment of Bt cotton provided a boost to biodiversity from a "marked increase in abundance" in beneficial insect predator ladybugs, lacewings and spiders as a result of the reduction in the use of pesticides.

One issue that anti-GM campaigners have half-right surrounds the spread of 'superweeds' that can tolerate a certain herbicide nearby herbicide-resistant GM crops. Weeds that are susceptible to the herbicides die off while those that are not continue to live. This is just natural selection and would happen anyway, but evolution has been sped up by farmers' over-reliance on a single weedkiller. It is indeed a serious problem, and resistant weeds have spread rapidly in the US, choking off production at a cost of millions in losses.

But this can be solved by a switch to a different herbicide or better crop rotation, with varied planting cycles, more temperate herbicide use and more locale-specific seeds. As these techniques vary season to season and year to year, resistance would evolve much more slowly.

Such a medley of tactics is just pretty elementary integrated pest management, but these practices have been forced out or forgotten as the supermarket chains pressure producers to grow as cheaply as possible.

Non-herbicidal solutions and such variability in planting practice will increase costs. Additionally, the likes of Monsanto and Syngenta make no money if a farmer plants 'cover crop'—a crop planted primarily for purposes of soil quality or fertility, pest and weed control or disease prevention. Meanwhile, agribusiness research is biased towards where the money is— newer, stronger herbicides. For these companies, superweeds are just another market opportunity.

The crucial point though is that over-reliance on certain products doesn't come from genetic modification, but the economics of the modern farming system.

Activists also regularly accuse GM of being 'linked to' the production of hectares upon hectares of monoculture crops, reducing biodiversity and contributing to soil erosion. Monocultural production can indeed be a problem, but this is an issue with non-GM monoculture as well. So criticise monoculture—and the economic relations that encourage its development—not GM.

By focussing on GM as the cause of superweeds and monoculture, campaigners are again letting the real villain—the free market—off the hook.

*

Don't forget that there are large corporations that have a great deal at stake here as well and are bankrolling many groups involved in the anti-GM fight. We have to acknowledge that the anti-GM position actually benefits a range of multinational corporations. In last year's California ballot initiative to try to legislate GM food labelling, the second-largest backer[96] of the Yes campaign, Mercola Health Resources, is a dietary supplement firm whose owner, Joseph Mercola, according to *Quackwatch*[97], has three times been warned by the FDA to stop making illegal claims about his products. Union-busting Whole Foods Market also backed the Yes side, while Just Label It, the national pressure group fighting for GMO labelling, was chaired by the head of Stonyfield Farm, an organic dairy firm that is 85 percent owned by French multinational Danone, simultaneously both the largest dairy product company and bottled water company in the world, whose 2010 revenues amounted to €19.3 billion.

The Cornucopia Institute, which was naming and shaming organic food companies whose parent firms backed the No side in California, was clear it is onside with Big Organic: "The Cornucopia Institute ... stresses that the organization is not against corporate involvement in organics. We welcome corporate involvement in the organic food industry, but only when the parent company subscribes to the values that the organic food movement is based on."

As the institute's consumer guide[98] to who was funding the two sides of the battle made graphically clear, they don't believe the antagonism is corporate behemoths vs family farmers, cooperatives and peasants; the division is 'chemical' vs 'natural'.

You say you are concerned about corporate control of the food industry? Well, you could actually be an accidental shill for a group of multinational corporations in this fight.

Remember how the biofuels industry initially was touted as a green alternative to fossil fuels by environmental groups, but when it was discovered that they were worse for the environment, biofuel producers pulled (and continue to pull) all manner of tricks to prevent their loss of subsidies. Could it not be that the organic industry is engaging in the same behaviour now?

Perhaps we could get all conspiratorial about the connections in the way that anti-GM website GM Watch regularly engages in accusations of guilt by association! We could note how last year's famous Seralini rat-cancer study was bankrolled by Auchan and Carrefour, the two French retail multinationals.[99] The latter supermarket chain launched an ad campaign for a range of GM-free products just five days after the publication of Seralini's paper.

Organic is big business these days. World organic food sales soared from $23 billion in 2002 to $52 billion in 2008, becoming the fastest growing sector of the American food marketplace, and as of last year, most independent organic food processors[100] in the US had been swallowed up by multinational firms. And one could imagine that they might not want their profitable new market endangered.

A 2012 meta-analysis[101] (basically a study of lots of studies — in this case 66 of them covering 34 crop species in both the developed and developing world) found that overall, average organic yields are 25% lower than conventional. So, in using up a third again the amount of land that conventional agriculture does, organic does not come off well in terms of land-use. This is not industry spin; it comes from researchers with McGill University's Land Use and the Global Environment Lab[102] in Montreal, a team of people very much dedicated to environmental preservation. (Not that the quality of a conclusion should be judged based on who it is doing the concluding. Bad research doesn't become good research when it's done by someone with good politics, and vice versa.)

If you were into Dan-Brown-style collusion and intrigue, you might be driven to remarking to yourself how convenient GM food scares, promoted by researchers funded by supermarket multinationals, are at a time when the evidence is beginning to show that organic food offers no additional nutrition, contains 'natural' pesticides that can be as toxic as synthetic ones, is less effective in preventing the spread of pathogens[103], and may actually be worse for the environment.

There are significant interests out there who stand to gain a lot from the continued mistaken belief that anything that has been genetically modified is inherently harmful.

In the 1960s, while the scientific consensus was that smoking was dangerous, Big Tobacco paid off unscrupulous scientists to keep alive the doubt in people's minds over whether there was any harm. Today, while the scientific consensus is that GM is safe, is Big Organic hiring unscrupulous scientists in order to keep alive the same sort of doubt, but doubt as to whether GM is safe? Merchants of doubt indeed.

Not that I would make any such remarks myself. I'm not the conspiratorial type. You might think that. I couldn't possibly comment.

While most people continue to believe that organic food is healthier than conventionally grown or raised food, meta-analyses have found no evidence that organic is more nutritious. It is also commonly believed that organic farmers use no pesticides, but this also is not true. Organic farmers use non-synthetic pesticides, but these remain chemical products that have the attributes of warding off or killing pests such as insects or weeds. Many 'natural' pesticides are not as effective as their synthetic cousins, and so farmers have to use more of them.[104] Just because they aren't synthetic does not make them safer. Those organic pesticides that have been studied have been found to be no less toxic than synthetic pesticides. Organic food is also damn expensive. An all-organic grocery cart is going to increase your weekly food bill by 49 percent.[105] Even putting the Great Recession aside, the income of 90 percent of Americans has been stagnant since the 1980s. At the bottom end, the benefit formula of the US Supplemental Nutrition Assistance Program (formerly Food Stamps) assumes that recipient families will spend 30 percent of their net income on food. Is it realistic to demand this sort of increase in spending on food? Even if organic food were demonstrably safer and more nutritious than conventional food, the difficulty of affording it, and the haughtiness and conde-scension of those who can toward those who cannot, would not be issues that any self-styled progressive should dismiss. the understandable desire on the part of moms to feed their kids the best food they can, combined with the guilt that they aren't always being the best parent they can be, is a gift to the multina-tionals behind the organic industry that is expected to be worth almost $200 billion by the end of the decade[106]—a gift that helps them get away with charging extortionate prices for staple food items.

I should add here that I don't want to beat up on organic too much. There are many other issues to consider than just yield and direct and indirect land-use, such as energy use, water use, water quality, nutrients, impact on biodiversity, carbon emissions, etc. For some crops and in some contexts, organic beats conventional.

In the end, what is going on here with opposition to genetic modification is the import into left-wing thinking of the logical fallacy of an 'appeal to nature'—the idea that what is found in nature is good and what is synthetic is bad.

Transferred to human ecology, the inherent conservatism of this should quickly be revealed: everything, or everyone—peasant, lord and king—has his place within the 'natural order'. It is a defence of the status quo against the 'unintended conse-quences' of social programmes by interventionist governments. How alike are the arguments against genetic engineering and 'social engineering'!

Let's uproot an unjust political economy, not GM crops.

*

Anti-GM activism is just the most high-profile example of this phenomenon of the appeal-to-nature fallacy on the crunchie left. Many of the same environmental groups that oppose GM also campaign against nanotechnology, stem cell research, cloning and the emerging field of synthetic biology. There will be perfectly reasonable ethical, environmental, economic and political concerns about not just these areas, but with regards to any new technology. But one has to be able to separate legitimate questions about the companies involved in the development of these areas from irrational opposition to anything new, simply because it sounds 'unnatural'.

This fallacious naturalistic stance reaches the height of irresponsibility with anti-vaccination campaigners and 'vaccine resisters' — parents who refuse to vaccinate their children, endangering the wider community via the deterioration of the herd immunity that widespread public vaccination engenders. Vaccines do not just protect an individual from a given disease, but once a large enough number of people in an area are vaccinated, there are fewer hosts and the chain of person-to-person infection is broken. This offers protection or 'herd immunity' to those immunocompromised individuals such as infants, chemotherapy patients, the elderly or anyone with an illness or condition that weakens their immune system, and so cannot be vaccinated. There are sects that refuse vaccination on religious grounds, but some of the most fervent opposition to vaccination comes from the same sort of crunchy, Whole-Foods-and-farmers'-market progressives who embrace alternative medicine in an entirely self-destructive and ignorant effort to outfox 'Big Pharma'. As a result of their efforts, in the last decade, we have seen a return in outbreaks of vaccine-preventable diseases such as measles, mumps, rubella and whooping cough throughout the West, with worryingly large clusters of vaccination refusal in middle-income liberal-left regions and cities where parents believe they are doing righteous battle with the sinister drug companies forcing vaccines on their little darlings.

And yet the reality is that vaccines are deeply unprofitable. So many pharmaceutical companies have abandoned not just vaccine research and development but production as well, that by 2003, the US began to experience shortages of most childhood vaccines. The situation is so dire that the CDC maintains a public website tracking current vaccine shortages and delays.[107]

As a journalist, I have repeatedly covered episodes of the dastardly crimes of the pharmaceutical industry, most recently tracking the discovery void of the last three decades since these companies got out of the business of developing new classes of antibiotics due to their great expense and low returns on investment (resulting in some 2 million Americans being infected with drug-resistant bacteria every year, according to the Centers for Disease Control, and 23,000 deaths[108]), their role in the deadly return of tuberculosis—one of our greatest killers, and their failure to develop treatments for Ebola despite having multiple candidate options ready to go if just someone, anyone, could pay for the clinical trials. I know these villains well.

But I also know the tremendous advances that evidence-based medicine has achieved over the last 200 years as a result of the germ theory of disease, sanitation, antibiotics, vaccines, pharmacology, lab technology and genetics. As Ben Goldacre, the doctor and health campaigner who manages to be simultaneously Britain's most trenchant critic of Big Pharma and of medical frauds such as homeopathy, herbal medicine, acupuncture and 'nutritionists', puts it: "Repeat after me: pharma being shit does not mean magic beans cure cancer." The socialist left, with its historic commitment to reason and science, has to separate itself from the distractions of the crunchy left.

We could do far worse in this regard than learning from the AIDS campaigners of the late 80s and early 90s in organisations like ACT-UP and the Treatment Action Group. They described and continue to describe themselves as "science-based treatment activists." While engaging in multiple high-profile acts of militant civil disobedience against the pharma giants and both Republican and Democrat politicians, they also soberly, rigorously plunged deeply into the science of their condition, and were willing to change tack upon the advent of new evidence, as happened when early demands of expanded access or "drugs into bodies," as was the slogan of the time, proved to be insufficiently nuanced. Despite most of the activists lacking any formal medical training, the extent of their evidence-focussed self-education and the quality of their reports and recommendations were such that clinicians began to recognise them as their equals in an understanding of the disease. And through this combination of a grounding in science and militant activism, ACT-UP and TAG changed the course of an epidemic, forcing governments to care about a plague killing queers, drug users and minorities.

Likewise, we can learn from South Africa's Treatment Action Campaign, with its direct action, roots in anti-apartheid and socialist activism, and fight for equal access to expensive HIV/AIDS medication against a neoliberal government. They too emphasised scientific rigor in the face of President Thabo Mbeki and his health minister, Manto Tshabalala-Msimang, who were both steeped in HIV-denialist pseudoscience[109], and throughout the 2000s took to recommending patients take vitamins, garlic and beetroot rather than antiretroviral drugs. The government's inaction resulted in delayed testing, prolonged illness, and an estimated 300,000 avoidable deaths[110]. TAC was shot by both sides, though, fighting not just the denialism that was official government policy, but also the irrationalism of community stigma and popular embrace of alternative-medicine snake oil, traditional healing and conspiracy theories. Traditional healers even protested outside TAC offices. Not shy of sharp criticisms of the ANC either, they have racked up successes, including forcing the government to make antiretroviral drugs available through public clinics, pressuring companies into reducing their prices, and winning the right of prisoners to access treatment. South Africa now has the largest treatment program in the world.

*

The anti-Vietnam War activist and heterodox economist Michael Albert years ago issued a jaunty, curmudgeonly aphorism that should be made into stickers to be slapped onto every anti-GM campaigner, crunchie alternative medicine evangelist, and anti-vaxx woo merchant's notebook, laptop and bike frame:

> There is nothing truthful, wise, humane, or strategic about confusing hostility to injustice and oppression, which is leftist, with hostility to science and rationality, which is nonsense.

The origins of this nonsense, this multifaceted antipathy to medicine, technology and progress as well, can be found in romanticist counterrevolutionary thought that emerged in the 18th Century in opposition to republican movements, as we'll see shortly.

It is a cuckoo's egg in the nest of the left.

9

Who's Afraid of 'Big Kit'?

A couple of years ago, I went to a pretty impressively organised activist variety show in London dedicated to alternative energy solutions, with art, bands, films, debate, book tables and a reasonable selection of beers. As is typical of these sort of events, it was a mixture of the genuinely inspiring, galvanising and moving—and the naïve, bonkers and profoundly mistaken. The programming included truly vital short documentaries exposing the grim ecological threat from Alberta's Tar Sands; the dash for oil and gas in the melting Arctic; and Shell's human rights abuses in the Niger Delta; and speakers talking about urban fuel poverty in Britain and the high-level corruption and pitiless repression that accompanies oil as it pipes its way westward from Central Asia. But the evening also included breathless celebrations of micro-energy projects in rural Indonesia and Brazil that can barely power an oven; an embrace of market-led responses to climate change such as carbon trading; and facile, blatantly commercial publicity for boutique ethical producers: "Saving the planet is as simple as switching to Ecotricity!"

One of the main speakers, Nnimmo Bassey, the former chair of Friends of the Earth International, issued what amounted to a call for UN military intervention to prevent environmental damage. He promoted legal recognition of 'ecocide' as an international crime against peace in the Rome Statute treaty that established the International Criminal Court in The Hague. Because this concept is couched in the same language of the emerging norm of 'Responsibility to Protect' (R2P) that buttresses such humanitarian interventions as the Iraq War, the invasion of Afghanistan or the Nato bombardment of the former Yugoslavia—the sort of military endeavours that most people in

the room at other times would unquestionably march against—
one would have thought the potential for imperialist shenanigans
embedded within the ecocide concept would have given them
pause. With ten countries having already adopted domestic
ecocide laws, and two states having enshrined Mother Earth
rights in their constitutions, this is hardly an idea from the legal
fringe. Australian political scientist Robyn Eckersley, who,
writing in the journal of the century-old New York-based
Carnegie Council for Ethics in International Affairs, contends
that "crimes against nature" are an exact parallel to the existing
norm of "crimes against humanity" and argues for the need to
extend humanitarian intervention to cover "ecological intervention," which would involve "the threat or use of force by a
state or coalition of states within the territory of another state *and
without the consent of that state* in order to prevent grave environmental damage" [emphasis in original].[111]

But what struck me as the most objectionable, because it is so
typical of the wholly inadequate small-is-beautiful approach to
climate change, was when one speaker at the event described
how the cornerstone of any solution to global warming has to be
the "decentralization of energy rather than Big Kit," mocking
large-scale infrastructure efforts as the quintessential colonialist,
masculinist, swaggering-big-dick exemplar of all that has gone
wrong since the Industrial Revolution.

This prejudice against Big Kit is pervasive on the green left
and I want to spend some time unpacking such ideas, and in
particular Naomi Klein's declaration that coal "extractivism" is a
"war on nature," her attack on China's "massively disruptive
mega-dams," and damnation of nuclear power as the supreme
"heavy industrial technology" emblematic of man's hubris.

One can certainly mount a strong argument for the former
case. A 2014 *Nature Climate Change* paper from climate modellers
Neil Swart and Andrew Weaver at my alma mater, the University
of Victoria in British Columbia, demonstrates the scale of the

threat from coal. The researchers found that if the truly massive amount of unconventional, bituminous oil from the Alberta tar sands were extracted—all 1.8 trillion barrels of oil-in-place[112]— its potential for global warming would amount to an additional 0.36 degrees Celsius. Remember that the internationally agreed target for an upper limit of warming is 2.0 degrees. As worrying as this is, the average surface temperature rise pales in comparison to that of coal. Due both to coal's large tonnage of feasible extraction and its high carbon content, total global carbon resources have a potential for warming of 14.79 degrees. Read that again: *almost 15 degrees.* This makes coal public enemy number one even if we leave aside the deaths from particulate emissions (24,000 annually in the US alone), acid rain, heavy metal pollution, fly ash spills and groundwater contamination.

Yet there's nothing intrinsically evil about coal, or for that matter oil or natural gas, however malevolent Klein's invented word, 'extractivist', sounds. Let me be really clear here: coal is *great.* Or, rather, coal *was* great. What made fossil fuels so fantastic was their tremendous energy density—the amount of energy stored per unit of volume—and, crucially, their porta- bility. For most of human history and prehistory, we primarily depended on biomass and the muscle power of humans and animals in order to do work. Wind and water power were also major sources of energy until the 19th Century, but both required methods of energy harnessing, windmills and water wheels, that are in general tied to geography. You can't just take a river and water-wheel with you wherever you go (although wind power harnessed via sails is indeed mobile, at least at sea, and only when the wind is blowing). Wood and charcoal were great though—all that energy bound up in logs that were easy to access and which you could take anywhere and use at any time, unlike the wind, which didn't always blow, and water, which didn't always flow at the desired speed. While coal pits were dug as early as the 13th Century, it was only in the 17th Century due to deforestation that people began to switch to the more difficult ways to access coal. But coal is about three times more energy-dense than wood, and it is at this point that we see the major discoveries such as the coal-fired steam engine that would propel the industrial revolution.

An easy way to understand how much of a leap forward this represents is to compare how long a 100-watt light-bulb could run on one kilogram of different fuels. One kilogram of wood can power a light bulb for just over a day, while coal can power it for just under four days, and oil about five days (crude 4.8 days; diesel 5.3). Moreover, even though the internal combustion engine invented in the late 19th Century was not particularly efficient, oil was extremely easy to transport and even easier to use than coal (pouring into a tank versus shovelling coal into a furnace). Even to this day, our very best non-rechargeable lithium batteries are about a tenth as energy-dense as wood, while lead-acid car batteries about a hundredth. (At the other end, uranium is *six orders of magnitude* more energy-dense than oil.) Indeed, one can sort of think of fossil fuels as the perfect battery. All the tremendous, brilliant advances of industrialisation have been fossil-fuel-powered. We could never have built any of the wonders of the modern world on the backs of horse power, water, wind or wood. That's why coal and oil and gas are great.

The trouble is that they are incredibly dirty—something we've long known in the case of soot particulate from wood and coal, the source of the great 19th Century pea-soup fogs of London and the great 21st Century (wonton soup?) smogs of Beijing—and are powerful greenhouse gases. We definitely need to stop using them now that we know about their role in climate change. But the requirement to switch away from fossil fuels, and coal above all, doesn't make them *evil*. And if we are at the very least going to maintain, let alone advance, humanity's current level of development, whatever suite of options we choose to replace fossil fuels will have to match them attribute for attribute—energy density, portability, lack of intermittency, abundance, ease of availability and ease of use—in what made them great.

*

China is the largest consumer of coal in the world. The energy source provides the Middle Kingdom some 70 percent of its electricity, and as a result, China has been since 2007 the world's largest emitter of greenhouse gases. In 2002, Chinese emissions represented roughly 14 percent of the global total. Since then, China's emissions have climbed 8 percent annually on average—a staggering increase—overtaking the US in 2006. As of 2013, China represents 28 percent of global emissions, has greater emissions than the US and EU combined, and has *per-capita* emissions 45 percent above the global average, emitting more than the EU on a per capita basis. Even taking into account historic emissions since the industrial revolution, the country remains the third largest emitter after the US and the EU.

Of course, to suggest that China forgo its breakneck growth for the sake of the climate is to deny it the right to develop its standard of living up to that of the West (again, acknowledging all the caveats I earlier mentioned about the meaningfulness of a 'Western standard of living'). And a fundamental principle of the climate justice movement is that the Global South should not have to sacrifice its development. The most radical climate activists accept that China, like the rest of the non-Western world, has a right to develop.

But it gets worse. Once built, coal-power plants have a lifetime of operation of around 40 years. This means that China has already locked in more CO_2 emissions than it has emitted up to now. And according to analysis from the Global Carbon Project—a group of climate researchers who track national emissions and pledges within the framework of international climate talks to mitigate emissions—in order to meet the internationally agreed limit of 2°C of warming of average global surface temperatures, China's carbon emissions would have to peak instead at some point between 2015 and 2020, and then decline roughly 8–10 percent per year afterward over the following 35 years.[113]

So to recap: China's emissions are growing 8 percent a year, but they have to be reducing their emissions by at least 8 percent a year. It is the most hairpin of u-turns. For comparison, the most successful decarbonisation programmes in history, the rapid transition to nuclear power by France, Sweden and Belgium in the 1970s and 1980s, with France now producing 78 percent of its electricity from this low-carbon source, enjoyed reductions in emissions of 4 percent a year over the course of roughly a decade.

And all this is to have a 50 percent chance of success (one chance in two, a toss of a coin) in avoiding greater than two degrees of warming. To up the likelihood of success to, say, 66 percent (two chances in three), the minimum mitigation rate jumps to 10.5 percent per year.

This is an absolutely elephantine undertaking, and throws into relief the debate around 'Big Kit'.

Can China do it? There is, actually, ample hope. In the last decade, China has become the world's largest producer of wind energy and second largest producer of solar energy. A total of 28 nuclear plants are under construction, atop the 20 currently in operation, giving China a nuclear sector growing faster than anywhere else in the world. The country hopes to boost its nuclear capacity to 88 gigawatts by the end of the decade, an increase of 80 percent, and 400–500 gigawatts by mid-century. For comparison, the Chinese electricity system alone *currently* produces 1,250 gigawatts of power. And China is also the country that builds a new skyscraper every five days, and has in the last few years constructed some 30 airports, 25 subway systems, the three longest bridges[114], the largest hydroelectric dam, and thinks nothing of flattening 700 mountains[115] to throw up a brand new metropolis for 500,000 residents from scratch.

*

Let's have a closer look at that hydroelectric leviathan. The country's Three Gorges Dam, at 60 stories tall and 2 km wide, is not just the largest hydroelectric plant ever built, but the largest *power plant* ever built. It is also the largest concrete structure in the world, requiring some 28 million cubic tonnes of the stuff, employing over 40,000 workers and taking 17 years to construct. The sheer mass of water in the gargantuan reservoir behind the dam is so vast that it even fractionally slowed down the rotation of the Earth. However one feels about the dam politically or environmentally, it is plainly one of the most gob-smackingly colossal objects man has wrought. Isambard Kingdom Brunel, that giant of Victorian civil engineering works, would tremble before it. It is a Sinitic Galactus the Devourer landed on Earth, a Death Star of concrete. It is *god-sized*.

The environmental transformation for which it is responsible is undeniable: massive landslides, upstream effluent bogs, industrial pollution produced by its construction as well as reduced aquatic biodiversity and a threat to one of the world's biggest fisheries. Dams in general are indeed disruptive. Some 60 percent of the world's largest rivers have been augmented by dams and reservoirs, both for electricity generation and irrigation purposes. All this damming has resulted in freshwater ecosystems shifting from some of the most biodiverse areas on the planet to some of the most threatened, by buggering up the thermal cycle of freshwater ecosystems; altering the sediment load of different parts of river systems; and the sinking of delta plains due to loss of sediment. This latter development in turn also increases the risk of flooding of cities built on those plains. Upstream, as will all dams, the large reservoirs submerge great tracts of land and all the terrestrial organisms that have adapted to suit such ecosystems, a process that is only exacerbated by habitat fragmentation—yet another driver of biodiversity loss.

Yet the Three Gorges Dam also massively reduces the country's prodigious coal consumption, preventing the emission of 100 million tonnes of greenhouse gases, according to its boosters in the country's National Development and Reform Commission. What will be the net ecological impact of the mega-dam and the 130 other dams in some stage of construction or planning across the Chinese southwest? How significantly, for example, will the methane released from rotting vegetation submerged in reservoirs mitigate the wider reductions in carbon emissions? Some estimates put the greenhouse gas emissions from methane over the average life of a dam to be equivalent to between a third and two thirds of that of a natural-gas fired power plant. It will be years before any objective net assessment can be mounted.

Nevertheless, hydroelectric power is responsible for 16 percent of the world's renewably produced electricity—easily outstripping wind, solar, wave, geothermal or other renewable sources. Also, unlike intermittent wind and solar, hydroelectric can generate power whenever needed. So despite the undeniable environmental harm that dams cause, for now at least, preventing runaway climate change must take precedence and that means that hydroelectric will have to continue to play a major role in any honest emissions-reduction strategy.

Moreover, if we were to scale up Klein's favoured options, wind and solar, to levels where we can genuinely say that fossil-fuel-based energy production is being replaced, then these sources too would be producing their own substantially scaled up environmental damage, whether from the steel production required for pylons, concrete for the bases and fiberglass (which cannot be manufactured without petrochemicals) for the blades, or the heavy metal pollution from solar panel manufacture.

According to the California Department of Toxic Substances

Control, solar firms in the state were responsible for the production of 21,000 tonnes of hazardous waste from 2007–11. The Silicon Valley Toxics Coalition, worried about how e-waste and solar photovoltaic waste is currently shipped to some of the poorest parts of the world—and to domestic prisons—for rudimentary manual disassembly and sorting, launched its Clean and Just Solar Industry initiative in 2009 to pressure for change in the sector.

Mathematical ecologist Robert Wilson has crunched the numbers and found that on average, one megawatt of wind capacity demands 103 tonnes of stainless steel, 402 tonnes of concrete, 6.8 tonnes of fibreglass, 3 tonnes of copper and 20 tonnes of cast iron. This involves scarring the Earth via open-cast iron ore mining, typically in Australia or Brazil. The Mount Whaleback iron ore mine in the Pilbara region of Western Australia owned by mining giant BHP Billiton, one of the largest such mines in the world, is five and a half kilometres long, two kilometres wide, and the company plans to keep mining there till they reach a depth of half a kilometre. From the air, the rust-red chasm, with its perfectly regular bull-dozed terraces, appears as a man-made Grand Canyon. The vast diesel-run machines and vehicles that extract and move the rock are the size of mansions, but even they are dwarfed by the ocean-faring, and again, diesel-fuelled bulk carriers that ship the mineral around the world. The ore is then smelted into steel, most typically using a blast furnace.

And now comes the real doozy of the whole process, when iron ore has its impurities and oxygen removed. That is, iron oxide must be transformed into iron, the by-product of which reaction is carbon dioxide. Iron ore, limestone and coke from coal go into the furnace, and out comes pure iron and CO_2. (Oh, and slag, a glass-like waste product mounded into heaps; the sort of slag heaps that my mum would ride over on her bicycle as a child just off the Mossley council estate in Bloxwich in the

West Midlands of the 1960s. She still has tiny fragments of slag embedded in her elbows and knees from when she skidded off her bike.) So in terms of climate change, the major problem is not even all the greenhouse gases emitted as a result of the machines engaged in extraction, the vehicles that transport the ore, or the fuel used to fire the process. It is the carbon emissions that are an *unavoidable part of the chemical process itself.* A similar hurdle is presented by cement production, which involves turning limestone (calcium carbonate) into lime (calcium oxide). Replacing the coal or gas used in the kiln firing isn't the biggest problem; again, it's the carbon dioxide that is the ineluctable by-product of this chemical reaction.

And were we to make a transition to 100 percent wind-powered electricity over the next 20 years (a back-of-the-envelope scenario Wilson uses for comparative purposes — obviously no one is suggesting we embrace just wind power), this switch would require roughly 50 million tonnes of steel, 200 million tonnes of concrete and 1.5 million tonnes of copper every year over that period.[116] We think of wind power being so clean and cuddly, but what is all of this, but Naomi Klein's 'extrac-tivism'?

It is true that the European Union has funded a handful of pilot projects to develop coke-free steel-making and a pioneering process involving a magnesium oxide-based alternative to cement whose production absorbs more CO_2 than is emitted.[117] I wish them the best of luck. Most other aspects of modern life have readily available climate-friendly alternatives so long as the political will is there to adopt them, but not yet cement and steel production.

None of this should be seen as an argument against wind or solar *per se.* (Indeed, with respect to solar, we are still very much in our infancy of developing the suite of technologies that can capture energy from the sun. Unlike wind, solar power research is pretty open-ended and exciting. Our best technologies cannot yet match the engineering wizardry of plant photosynthesis, so there is still a lot to play for there. And the possibilities presented by space-based solar power (SBSP) and associated development of long-range wireless power transmission are equally exciting because in space, the available solar energy is ten times what we can access on Earth, and as Japanese Aerospace Exploration Agency (JAXA) researcher Susumu Sasaki puts it, "It's always sunny in space," which means that space solar can deliver baseload power. Space solar isn't even science fiction: JAXA is currently engaged in a basic research phase for an orbital solar farm that they hope to develop in the 2020s and commercialise into a full-scale space-based power station by the 2030s.) Rather, the point is that whichever options for energy we choose, we have to recognise that they will all involve some significant alteration of the environment. Steel isn't just necessary for windmill towers, but for hydroelectric turbines, transmission lines, geothermal wells, and the construction of all those electric trains and buses. The question that arises—and it is not an easy one to answer—is which options are optimal with respect to the maintenance of an environment maximally equipped to sustain limitless human flourishing, not which ones are extractivism-free.

Jensen, Kunstler, Kingsnorth and friends would thus respond: "Aha! So there really is no way for human civilisation to be compatible with the environment!" But here's the genuinely depressing news, the real catastrophe: we must acknowledge that there may well already be sufficient human-produced greenhouse gas emissions since the Industrial Revolution that runaway climate change is already on the cards. A vast chunk of the mammoth West Antarctic ice sheet has begun to disintegrate, it was reported in May 2014 by two separate groups of scientists in the journals *Science* and *Geophysical Research Letters*, a process that they conclude is now irreversible, which means that a sea-level rise of at least 10 metres is no longer a possibility, but guaranteed in the next couple of centuries.

This development in Antarctica is just one of many possible examples that shows how *withdrawing from civilisation would not stop the disaster from arriving*. It will require significant ingenuity to engineer a reverse of the processes we have inadvertently set in motion, likely even involving some way to produce a carbon-negative economy for a period. This will involve developing some technologies and processes that we do not really have yet. As yet unimagined new materials to replace steel and concrete are only two innovations amongst the hundreds that we will need. Better battery and energy storage technologies are high on the list as well, not least reducing the heavy carbon footprint of lithium batteries the production of which represents almost half the lifetime carbon emissions of electric cars. All of this will only come from the most advanced research laboratories and factories. By turning its back on the possibility of such technologies, on the very idea of progress, green anti-modernism *actually commits us to catastrophic climate change*.

Retreat from our predicament is not an option. We must push through the Anthropocene, indeed *accelerate* our modernity, and accept our species' dominion over the Earth.

*

It might also be useful here to make a distinction between dam-building in China and dam-building in other more democratic regions. The Three Gorges project involved the displacement of 1.2 million people from 13 cities, 140 towns and 1350 villages, according to International Rivers, an NGO that campaigns against hydroelectric power. Under China's autocratic system, these millions were pushed aside with many never receiving proper compensation while opponents were subject to police beatings and intimidation. Funds from the resettlement budget intended for household compensation were funnelled into projects like hotels and roads. It has been an embezzlement bonanza for local officials.

But, again, returning to Klein's concerns about scale and hubris, it should be clear that it is not size that is responsible for the human rights abuses and corruption, but the political system.

So long as local communities are genuinely involved in planning and implementation, there is no reason due to size or hubris that should prevent this as an option. We can imagine other ways of building hydroelectric plants, and point to Roosevelt's New Deal development of the Tennessee Valley Authority in the 1930s or the public-sector construction of some of the most ambitious hydroelectric construction projects in the world in British Columbia and Quebec from the 1940s to the 1970s. In both the American and Canadian cases, while there was some displacement and, again, environmental consequences from wilderness disturbance, at the time, the left viewed such projects as undertakings that expanded social justice, delivering cheap, reliable electricity to millions in communities that previously had none. Meanwhile the right despised the TVA for its socialistic taint.

Furthermore, dams in areas with greater levels of precipitation, for example, are not as destructive as those in arid zones, so appropriate siting of a dam has a role to play as well. We can also learn from our mistakes and build better dams, or modify existing ones so that they begin to more closely simulate a river's normal changes in flow, temperature and sediment load, thus reducing the impact on riverine ecosystems. At the same time, while shifting to a *complete* mimicry of a river's flow, without vast reservoirs—a form of hydroelectric power generation called 'run-of-the-river' (ROR hydro)—may be an appropriate option in some geographical areas, it should be kept in mind that such options, just like wind and solar, involve significant intermittency of the electricity produced, as they are sensitive to variations in seasonal river flows. This excludes them as options for delivery of baseload power—the dependable, core generation that can consistently meet primary electricity demand.

Sadly, instead of the large-scale, democratically driven, public-sector electrification works that will deliver cheap, reliable, abundant power to all communities, development NGOs these days favour micro-energy projects featuring off-grid water-wheels and wooden wind turbines that can produce just enough electricity to light a school house, or do-it-yourself biogas generators and solar food-dehydrators made from beer cans. "So that small communities can have sovereignty over their energy production," as an activist from the Brazilian landless workers' movement, the *Movimento Sem Terra* (MST), said in a video produced about their wind-turbine school in Itapeva. In July, 2014, Greenpeace installed a solar/battery microgrid as a business in Dharnai, a village in eastern India. The microgrid was to supply a miserly maximum of 100 kilowatt hours per year. But just a few months later, a government-owned energy utility extended the national grid to the area, offering unlimited electricity at drastically cheaper prices and undercutting the Greenpeace operation. Like the green NGO, private microgrid companies such as Husk Power Systems and Mera Gao Power, who also hope to make a killing from impoverished communities in Haiti and Nepal, are fuming that the public sector is "crowding out" their for-profit ventures. Nikhil Jaisinghani, of the USAID-grant funded Mera Gao Power, told the Bloomberg newswire of his company's preference for a more purely free-market environment without government subsidies: "In India, we can charge $2 a month, whereas we can charge $8 a month in Haiti."[118] Energy analyst Alex Trembath notes that analysis from the International Energy Agency found that the levelised cost for off-grid or mini-grid solar PV is just over $300 per MWh, and micro-wind about $250, while on-grid large-scale hydro costs about $50. Upon encountering the jealous butthurt of these NGO and private-sector micro-energy hustlers, Trembath joked: "[It's] a bit like saying that a new hospital came in and undercut Doctors Without Borders."[119]

One could weep at the poverty of aspiration here compared to the continent-straddling ambition of the *galacticos* of the old left. "Communism equals Soviet power plus the electrification of the entire country," declared Lenin. Once upon a time, bountiful, public, cheap energy was a key demand of progressives. Let's remember that as of 2014, according to the World Energy Outlook, it is still the case that 1.3 billion people around the world lack access to electricity and roughly 3 billion still cook their meals over open fires using charcoal, dung or wood as their fuel source.[120] Science journalist Fred Pearce wrote about the fitting of 300,000 Kenyan homes with rooftop solar panels in a 2014 article in the *New Scientist* on the current debate between traditional environmentalists and a growing number of 'eco-modernists' who favour public-sector 'Big Kit' clean energy projects: "A couple of panels on the roof can charge phones and run a few lights and a radio but would be no good for anything more demanding, like boiling a kettle. Most Kenyans would probably prefer to be hooked up to centralised power, but the grid only reaches one-fifth of the country." (To be fair, Pearce says that the eco-modernists' defence of hydroelectric is somewhat wanting too, referencing an analysis of 245 large hydroelectric projects built between 1934 and 2007 by Bent Flyvbjerg of the University of Oxford's Saïd Business School: "It concluded that dams are mostly financial millstones: completed years late, almost 100 per cent over budget, and delivering less economic return than they cost to build. Recent dams are no better than older ones, and the bigger they are, the worse they perform." I quite agree. Just as small for the sake of small won't work, big for the sake of big won't either. Any decision on energy transition projects must be made on the basis of evidence, not the romantic modesty of the small, or the thrill and glamour of the colossal.)[121]

I do not want to denigrate the amazing work of the MST, a militant, 1.5-million-member strong land-reform social movement that engages in mass occupations of unproductive rural land. The MST has in the face of at times brutal repression successfully fought back against wretched injustice in a country where just three percent of the population own two thirds of the arable land. But if localist micro-energy projects are not a stepping stone to the public-sector-orchestrated electrification of the country, and instead an end in themselves, then they are more likely to serve than to challenge elite interests, as they are a retreat from the historic goal of capturing the state. To deliver on the promise of social justice, we need a high-energy planet, not modesty, humility and simple living.

This small-is-beautiful attitude also dove-tails so perfectly with neo-liberal hatred for the public sector. While some micro-energy projects co-operatively developed, elsewhere they are established by micro-enterprises and funded by micro-credit schemes that drive borrowers into debt traps or by remittances home from immigrants in developed nations.

And almost everywhere, much of the establishment green left wants energy to be made *more* expensive whether through the market-based mechanism of carbon trading schemes or imposition of carbon taxes. Green parties across Europe—who have for years played a kingmaker-like role in the European Parliament and have governed in coalition with parties of both left and right in Germany, Finland, Ireland, France and the Czech Republic—have been some of the biggest cheerleaders for energy deregulation and privatisation.

The Greens believe that only through full energy liberalisation can small, alternative energy producers get a foot in the door. Yet the sheer scale of transnational infrastructural investment for offshore wind, Saharan solar arrays and transmission across the Mediterranean, as well as new smart load-balancing grid—let alone the vast sums involved in the construction of nuclear plants—really are best served by a public sector that need not be restricted by the risk-aversion of market actors and which can borrow at much cheaper rates than the private sector. The gargantuan investment effort that will be required to shift to a genuinely clean-energy economy on the continent can only realistically be made by a Europe-wide public energy service. And with energy back in the public sector, once a democratic decision has been made by the European people to make such a shift, it could happen tomorrow, without having to invent Byzantine policies such as the boondoggle of the Emissions Trading Scheme to try to incentivise private companies into doing the right thing.

The same public-sector-led infrastructure argument could be made about the electrification of the transport fleet and extension of public transit systems and rail networks. All this is best performed by the public sector, but the Greens on the continent have been some of the biggest cheerleaders for rail liberalisation and in Ireland backed cuts to bus services. Whichever country in which it happens, privatisation of bus companies shrinks service on all but the most profitable runs. Rural public transport has steadily vanished in the neoliberal era. Charming, romantic and climate-friendly—but only thinly profitable—night-train services across Europe are disappearing as well, another cheery product of rail companies that are now in the business of delivering dividends to share-holders, not passengers to their destinations. There are certainly individual Greens who lament the waning of support for rail services, but this is in contradiction to Green Party's overall zeal for liberalisation.

Graham Greene, the left-wing-Catholic author of *Stamboul Train*—the classic novel of intrigue aboard a multi-day Orient Express train journey from Ostend to Istanbul—would surely disapprove of the party that bears his name.

*

The Left Defence of Nuclear Power

There is of course an upper bound to how much hydroelectric can offer, but that comes less from riverine impacts than from site saturation. That is to say, there are only so many places one can put a dam. This means that we'll need something else as well.

So, in the infrastructural spirit, how about we interrogate Klein's cheerleading of Germany's *Energiewende*—every green activist's favoured package of policies, a combination of feed-in tariffs for renewable energy and abandonment of nuclear power—that ostensibly demonstrates how any country can rapidly shift to a renewable energy-based economy, shall we?

In *This Changes Everything*, Klein enthuses how under the *Energiewende* ('energy turn' or 'energy shift') some 30 percent of electricity produced there now comes from renewable sources and could get to 50 or 60 percent by 2030 "mostly through decentralized, community-controlled ventures of various kinds, including hundreds of energy co-ops."[122]

Unfortunately the *Energiewende* is another example of activists so wanting to believe in something that all evidence to the contrary is dismissed as coming from climate deniers or sockpuppets of the fossil fuel industry. The reality is that many people such as myself who do genuinely accept the frightening reality of anthropogenic global warming and support a rapid shift to low-carbon energy also consider the *Energiewende* to be a neoliberal boondoggle that, according to the evidence from such neutral sources as the country's Federal Association of the Energy and Water Industry (*Bundesverband der Energie- und Wasserwirtschaft* – BDEW)[123] is actually pushing Germany's greenhouse gas emissions *higher*, climbing 1.6 percent in 2012 and 1.2 percent in 2013. In March 2014, the vice president of the country's Federal Environment Agency (UBA), Thomas

Holzmann, warned: "It is worrisome that the trend towards coal-generated electricity became even more pronounced in 2013. If it continues, we can hardly expect to achieve the Federal Government's climate protection goal for 2020."[124]

The proportion of electricity produced from renewable sources has crept up in the country, but these small gains have largely been counterbalanced by the necessary increase in coal-fired production. The reason is fairly straightforward: Germany is a northern country where the sun doesn't always shine and, onshore, the air is often calm, not blustery. The year 2013, for example, was an exceptionally windless one. Winter is a particularly grim time for solar power. This makes the energy produced from these sources highly intermittent. (One should also note that in Germany, as in the truly solar-ready Sahara, there is a daily phenomenon known as 'night' that adds to this intermittency.)

So this means that half the increase in renewables in 2012 actually came from hydroelectric, biomass and waste incinerators, because they are much more dependable. But there isn't much space left for extra hydroelectric in Germany, and the amount of waste than can be incinerated also has its limits. (And one would think that even with the advances in incineration technology reducing the amount of toxic fly ash, dioxin and furan emissions, it's hardly a green-energy solution activists would be trumpeting.[125]) Still worse, a full 28 percent of Germany's renewable energy production comes from the burning of biomass—mostly wood chips and energy crops. This flies in the face of environmental activists' absolutely correct fight against biofuels, due to their higher full life-cycle greenhouse gas emissions than fossil fuels and contribution to steep rises in food prices.

One day, we may have a major breakthrough in battery technology that will resolve some of this, allowing us to store excess capacity so that it is dispatchable when needed. But we're not there yet. Some researchers worry we may never get there. This is because batteries or other forms of energy storage require an expense of energy to make them. For an energy project to be worthwhile, you need to get more energy out than you put in. If the combined energy cost of building solar or wind projects and manufacturing batteries or energy storage—such as using excess electricity to pumping water up to a reservoir and then, when needed, recapturing that stored potential energy via hydro-electric turbines as the water rushes back down—is more than the energy put in, it's a waste of time. This sort of energy accounting is referred to as EROEI, or 'energy returned on energy invested'. Batteries or storage (both referred to as 'buffering' in the trade) significantly reduce the EROEI. But it's not good enough to just break even. At this point, you just end up back where you started. There has to be a pretty decent return, an EROEI threshold to be able to do the sort of things a society wants to do. A simple agricultural society would need a return of five to one. An advanced society involving healthcare, education and arts and scientific enterprise needs an EROEI of 14 to one. Disconcertingly, a series of EROEI studies appearing in reputable journals in 2013 and 2014 suggests that once all this is taken into account, even with the least energy-intensive form of buffering (pumped hydroelectric), photovoltaic solar clocks in well below this threshold, as do both onshore and offshore wind. In one such EROEI assessment by physicist Daniel Weissbach of Berlin's Institute for Solid-State Nuclear Physics and his colleagues, desert solar thermal power just about passes the test, and gas and coal pass it with flying colours (28:1 and 30:1[126]), but it's hydro and nuclear that come top of the class, with respective buffered EROEI scores of 35 to one and 75 to one. Other researchers contest these numbers because as simple as the concept of EROEI sounds, researchers use a wide variety of very different method-ologies for its calculation and never say never when it comes to battery innovation. But criticisms of the energy absurdity of

Alberta's tar sands often depend on similar EROEI studies, so what is good for the fossil-fuel goose is good for the photovoltaic gander, and if more and more studies begin to tilt in this direction, such findings should be deeply sobering for renewables advocates.[127]

Where genuinely clean nuclear power until the mid-2000s represented the largest proportion of Germany's electricity generation, this crown is now worn by brown coal (lignite), the least efficient and dirtiest kind of coal there is. Even as the country is mothballing its fleet of nuclear power plants, it is commissioning and opening brand new coal-fired power plants. The reason is due to the aforementioned intermittency of wind and solar, as the country's 'superminister' of economics and energy, Sigmar Gabriel, told the *Bild am Sontag* weekly: "You can't abandon nuclear power and coal at the same time."[128]

Even the idea that the *Energiewende* is some revival of national planning, the sort of public sector build-out of new energy and transport infrastructure that we do need to solve the climate crisis, is laughable. Rather, it is neoliberal business as usual, with the feed-in tariff offering subsidies to property owners and companies to install the solar panels and wind turbines. This subsidy is expensive, with the cost passed on as a surcharge to consumers, making electricity in Germany the most expensive in Europe. The average household now pays an extra €260 a year, while 2,300 energy-intensive companies are exempt from paying the surcharge—and costs are only expected to rise. From 2008 to 2011, the proportion of German households suffering from energy poverty climbed from 13.8 to 17 percent (a household is considered to be energy-poor when it spends more than 10 percent of its income on energy). In August 2013, a sparkling, gold-plated electrical cable and plug adorned *Der Spiegel*'s cover, with its feature article entitled: "Germany's energy poverty: How energy became a luxury good." The article quotes Stefan Becker, of the Catholic charity Caritas, saying that the situation is crippling for low-income households: "People here have to decide between spending money on an expensive energy-saving bulb or a hot meal." He fights to prevent his clients from having their electricity cut off after they fail to pay bills. "After sending out a few warning notices, the power company typically sends someone to the apartment to shut off the power — leaving the customers with no functioning refrigerator, stove or bathroom fan. Unless they happen to have a camping stove, they can't even boil water for a cup of tea. It's like living in the Stone Age."[129]

Some 300,000 households a year now are having their power cut off, according to aid organisations, and electricity termination warning notices were issued to 5.7 million households in 2013. Particularly risk-prone are those households with multiple children, the elderly, or those in need of care.[130] Indeed, the *Energiewende* is in effect a method of transferring wealth from the poor to the better off; renewable energy subsidies redistribute money from the poor to the more affluent, as happens when someone renting an apartment subsidises the landlord's mounting of solar panels on his roof via his electricity bill; and to large multinationals like Siemens, ThyssenKrupp and Volkswagon when they are freed from paying the extra bills in order to keep them internationally competitive. The situation is utterly perverse.

Meanwhile, the solar installation rate has actually collapsed by 70 percent over the last two years and Merkel has essentially given up on the piddling amount of solar and onshore wind production, shifting policy instead to more reliable offshore production[131]—not exactly the sort of vast steel-, concrete- and rare-earths-intensive apparatus that small-scale local energy co-ops are capable of constructing. The social-democrat minister Gabriel has come to the conclusion that the haphazard strategy is a mess. "Anarchy prevails in some areas. Everyone is joining in, but no one knows what direction to go in," he says. "I think we need to start over."

While *The Economist* laments the *Energiewende* as a "quasi-planned economy,"[132] again, the reality is that it is a direct product of the European Union's deregulation and liberalisation of the electricity market in the 1990s and 2000s, a series of regulations and directives that in effect 'Enron-ified' Europe. The Greens in the European Parliament, dominated by their largely neo-liberal German contingent, were far and away the biggest boosters of liberalisation, more so even than centre-right Christian Democrats and their allies, arguing that it would prevent large utilities from using their heft to arrest the entry of smaller competing companies producing green or cheaper power, and thus generate more investment in renewable energy. Instead, liberalisation has replaced democratically controlled public monopolies with unaccountable regional oligopolies whose profits are larger than some member states' entire GDP, and resulted in higher prices, profiteering and decline in network infrastructure and reliability. The Greens even backed the removal of energy regulators from democratic control, and criticised the European Commission for its timidity, insisting that the unbundling of production from transmission had not gone far enough.

No, true planning, the kind that will actually take us most of the way to a low-carbon economy, would involve taking the power companies back into true public ownership—that is, transforming them back into the providers of public services rather than operating as *de facto* multinational corporations. Rather than depending on arcane market mechanisms riddled with loopholes, a government could simply, democratically order, say, a European Electricity Service, to decarbonise, as the French government did in the 1970s when it decarbonised 78 percent of its electricity in just 13 years via a vast build-out of 54 nuclear power plants.[133]

The truth is that we can do this, we can solve the climate crisis. The solutions are at hand, but it will involve overcoming our irrational fear of nuclear power, and a genuine turn away from market-based solutions—an *Energiewende* of a red rather than a green type—toward planning, the public sector, and large-scale infrastructure, with the biggest role reserved for nuclear power.

*

"Nuclear is a heavy industrial technology, based on extraction," avers Naomi Klein, who goes on in *This Changes Everything* to quote psychiatrist Robert Jay Lifton: "[N]o technology does more to confirm the notion that man has tamed nature than the ability to split the atom."

Nuclear remains a nemesis of much of the green left. But we now know that background radiation is not the bogeyman some once thought it to be. Exposure to cosmic rays while taking two transatlantic flights (0.16 millisieverts of radiation), is roughly equivalent to the annual exposure of a UK nuclear power station worker (0.18 mSv), which is far less than the annual dose of the average US citizen from all sources (2.7 mSv), or exposure to radiation as a result of one CT chest scan (6.6 mSv) or the average annual dose from radon from the ground experienced by people who live in Cornwall (7.8 mSv). We also know that the new generation of dramatically safer reactors employing passive-safety systems *physically cannot* melt down, and that safe methods of waste disposal are proven. The amount of waste produced is also tiny compared to that of many other industrial processes, and far less hazardous. Radioactivity also decreases with time, but the danger presented by solar panel production, such as cadmium, mercury and lead pollutants, never goes away. Instead these pollutants bioaccumulate (there is ever greater concentration of the pollutant in an organism) and biomagnify (there is ever greater concentration of the polluting as you move up the food chain). Advanced nuclear power systems can completely recycle used nuclear fuel, actually producing a net positive balance of energy in this process. In a 2014 survey of all energy sources exploring which delivered the least direct harm to biodiversity, nuclear was amongst the best options, due to its small land and mining footprint.[134]

And the trump card is that nuclear has by far the best safety record of any energy source, clocking in at 0.04 deaths per terawatt hour, compared to wind's 0.15 deaths, solar's 0.44 deaths, hydroelectric's 1.4 deaths, oil's 36 deaths and coal's 100 deaths.[135]

All of this is to say that each of the three main arguments that the anti-nuclear movement mounts against this technology — meltdown, background radiation and toxic waste — are no longer the problems of a few decades ago, and perhaps were never the issues they were presented as. To complain about them in 2015 is akin to complaining about how annoying it is to have to rewind VHS cassettes in 2015.

*

Once built, nuclear is cheap. A 2010 study from the Organisation for Economic Cooperation and Development (OECD) assuming carbon pricing of US$30 per tonne (the commonly cited minimum price of CO_2 that would make renewable energy sources cost-competitive with fossil fuels[136]) estimated the cost of electricity from nuclear at the low end to be 4.87¢ per kilowatt hour compared to 21.93¢ for photovoltaic solar, and at the high end 7.74 ¢/kWh and 33.27 ¢/kWh respectively.[137] But reactors are stupendously expensive to build. Some 70–80 percent of the cost of electricity from nuclear comes from the up-front construction costs. And it can take decades for that investment to pay itself off, which makes the private sector extremely reluctant to go anywhere near nuclear without significant public subsidies. In the US in the 1970s, the financial standing of private companies that were involved in the American build-out deteriorated largely as a result of the mammoth construction cost overruns and had to hand over significant authority to public utility commissions. Why risk going to all this trouble when other energy sources are so much cheaper for similar rates of return on investment?

The key to France's nuclear success in the 70s and 80s, with a mean construction time per plant of just 76 months (just over six years), was making the transition largely prior to the advent of neoliberalism in the 1980s. Because of France's remarkable and rapid success, numerous energy policy analysts have investigated the reasons. Consistently, more than any other factors, researchers[138] have found that the essential element was France's muscular *étatist* institutional framework that permitted centralised decision making: a public-sector electricity monopoly enjoying substantial, concentrated engineering resources (the state engineers of the *Corps d'État*), and strong political support. Alongside this was an influential public research and development agency, economies of scale from standardised and large-scale build-out. All this is necessary, just as in any other *grands programmes* of state such as the Manhattan Project or the Apollo Program, due to high levels of technological risk, reliance on complex and large-scale equipment that requires a lengthy lead time and complex learning, and, above all, large amounts of capital.

"Nuclear power, perhaps more than any other industrial technology, is a child of state interests," wrote sociologist and historian of the nuclear industry, James Jasper in his exploration of the French experience. Since then, the French nuclear sector has been pummelled by a series of external shocks, the main one being the liberalisation of the European electricity sector, according to these same researchers.[139] The mighty institutional know-how of *Energie de France* (EDF), and indeed of the industry as a whole, has atrophied. In the late 1990s, EDF built a further four reactors, at roughly double the cost, adjusted for inflation, that France had enjoyed in the 1970s, a product of this deterioration of such preciously acquired competence, and of the liberalisation and out-sourcing of state processes.

All this is precisely why the strongest backers of nuclear power advocate that it cannot be anything other than strongly public-sector driven, and some of the fiercest opponents of nuclear power, beyond the usual green NGO suspects, are, surprisingly, actually on the free-market right. Because the very high up-front construction costs require companies today to depend on government loan guarantees, and the risk of default on these loans is high due to cost overruns, it leaves the state to pick up the tab should the loans turn bad. As a result, the libertarian Cato Institute has taken to warning against the "risky business" of nuclear power and both the American Enterprise Institute and *The Economist* recommend that nuclear be considered a "sideshow" that should only ever play a minor role in the energy mix. The scale of public investment and control required means nuclear smells too much like socialism to them.

The public-sector argument around nuclear power is essentially the same as that of Mariana Mazzucato, mentioned earlier in this book, and her argument that the private sector is in fact a timid, conservative economic actor in comparison to the state. (None of this is to say that to be leftist is perforce to be dirigiste. The trepidation that anarchists, libertarians and, yes, even liberals experience with regard to the state should not be dismissed out of hand. The state has not withered away under neoliberalism. State spending as a proportion of GDP has for the most part actually increased since the time of Thatcher and Reagan. Instead we are witness to the arrival of the *neoliberal state*, as British left-wing commentator Richard Seymour has put it, "reducing their democratic capacities by outsourcing previously accountable functions to businesses, quangos and unelected bodies." From the galloping advance of the security and surveillance apparatus, to a sort of dictatorship of the quotidian that bans smoking on patios, pens in peaceful protesters to 'designated protest zones', and arrests people for making joking on Twitter about airport security, neoliberalism is no enemy of the state.[140])

And the release of energy from the nuclear power we know, nuclear fission, or the splitting of large atoms such as uranium or plutonium into smaller atoms, is about three to four times less than that of nuclear fusion, or the combining together of the nuclei of lighter elements such as hydrogen or helium to form a heavier nucleus—the process that powers the Sun. The fuel is abundant—deuterium from seawater—and the fusion produces far fewer radioactive by-products because we are dealing with much lighter atoms. Nonetheless, there remain daunting engineering hurdles before fusion, often referred to as 'putting the Sun in a box', is a viable alternative, and we are some years or even decades away from commercialisation, but here again, it is the public sector that is taking the lead in the development of this new, almost limitless power source. The pioneering EU-led ITER (International Thermonuclear Experimental Reactor) project mounted in partnership with China, Japan, Korea, India, Russia and the US began construction in 2013 of its experimental tokamak reactor. At a cost of €16 billion, with considerable cost over-runs already, ITER is, unsurprisingly, a government funded project. (The deindustrialisation partisans of Greenpeace and European Green parties "deplore" the spending on fusion research, of course, claiming completely erroneously that it "would create a serious waste problem, would emit large amounts of radioactive material and could be used to produce materials for nuclear weapons.")

Nuclear power is thus, in a way, a good example of how it is indeed a case, in Naomi Klein's language, of the climate vs capitalism, but not in the fashion that she thinks. Rather, this is the case because it will take a substantial, global reversal of neoliberalism and an embrace of a strong, *democratic* public-sector ethos for us to take greater advantage of what is absolutely the strongest weapon we have in our arsenal against global warming.

10

The Dictatorship of the Expert-Ariat

Shortly after the 2009 Copenhagen UN climate talks collapsed, James Lovelock, a godfather of modern environmentalism and the founder of the controversial Gaia hypothesis, was asked by *Guardian* reporter Leo Hickman what should be done in light of the failure. Lovelock issued a call for what can only be described as a climate dictatorship.[141]

Rejecting the idea that a solution to climate change could be achieved in a modern democracy, Lovelock thundered that what was needed instead was "a more authoritative world" where there are "a few people with authority who you trust who are running it":

> What's the alternative to democracy? There isn't one. But even the best democracies agree that when a major war approaches, democracy must be put on hold for the time being. I have a feeling that climate change may be an issue as severe as a war. It may be necessary to put democracy on hold for a while.[142]

This call for a sort of benevolent dictatorship of science is increasingly being made for a range of problems that we confront globally, from climate change to biodiversity loss to antibiotic resistance. The willingness of so many progressives to—even temporarily—put replace democracy with the rule of 'experts' is a worrying development indeed.

*

Antibiotic resistance has become such a danger to public health worldwide, and government action has been so indolent and inadequate, that a pair of leading scientists impatient with the situation have called for a new executive global body to assume control of the problem. They want an international organisation similar to those currently tasked with navigating our species' response to climate change—basically an Intergovernmental Panel on Climate Change (IPCC), but for bugs and drugs and with more executive oomph.

Given the magnitude of the danger—the "apocalyptic" scenario, according to Sally Davies, the UK's chief medical officer, is that within twenty years we will completely run out of effective drugs against routine infections—it may seem a trivial, even irresponsible, exercise to fret over the democratic ramifications of such a move.[143]

However, considering how often this kind of technocratic proposal is the default response to any new scientific problem of profound import, democrats do need to consider whether other approaches are more desirable.

"So far, the international response has been feeble," wrote Jeremy Farrar, the director of the Wellcome Trust, the UK's largest medical research charity, and Mark Woolhouse, University of Edinburgh professor of infectious disease epidemiology, in a tub-thumping commentary published in the scientific journal *Nature* in 2014 and presented at a press conference at the Royal Society (giving the proposal the imprimatur of the august scientific body).[144]

The commentary took aim at the World Health Organization in particular, which in April issued its first ever report tracking antimicrobial resistance worldwide, finding "alarming levels" of bacterial resistance. "This serious threat is no longer a prediction for the future, it is happening right now in every region of the world and has the potential to affect anyone, of any age, in any country," the authors warned.

Despite the acceleration of this universal risk, the UN body responded by simply calling for better surveillance. "The WHO missed the opportunity to provide leadership on what is urgently needed to really make a difference," the authors wrote, acknowledging that surveillance is vital, but radically insufficient.

The growing threat from what are popularly termed "superbugs" is similar to that posed by climate change—they are "a natural process exacerbated by human activity and the actions of one country can have global ramifications," according to a parallel statement put out by the two authors' organisations.

They are not the only researchers or clinicians that have made the comparison between drug resistance and climate change. Last year, Davies described the situation as a more dangerous risk than terrorism, and a greater threat to humanity than global warming, telling the BBC, "If we don't take action, then we may all be back in an almost nineteenth century environment where infections kill us as a result of routine operations."[145]

So many medical techniques and interventions introduced since the 1940s depend on a foundation of antimicrobial protection. The gains in life expectancy that humanity has experienced over this time depended on many things, but they would have been impossible without antibiotics. Prior to the development of antibiotics, bacterial infections were one of the most common causes of death.

We need to keep discovering new classes of antibiotics because over time the bugs that are susceptible to the drugs are eradicated. Those with random mutations that make them resistant survive, reproduce and eventually dominate. This is just evolution.

And yet for almost three decades, there has been a "discovery void."[146] No new class of antibiotics has been developed since the use of lipopeptides in 1987. The reason for this is straightforward: big pharmaceutical companies have refused to engage in research into new families of antibiotic because such drugs are not merely unprofitable, but are antithetical to capitalism's operating principles. The less they are used, the more effective they are.

As these firms readily admit, it makes no sense for them to invest an estimated $870 million per drug approved by regulators on a product that people only use a handful of times in their life, compared to investing the same amount on the development of highly profitable drugs that patients have to take every day for the rest of their lives.

Some governments have begun to partially recognise this market failure. The European Commission has set aside €600 million for an "innovative medicines" program endearingly named "New Drugs 4 Bad Bugs."[147] But the scale of investment allotted by governments to this solution remains inadequate.

Hence Farrar and Woolhouse's demand for the establishment of a global, scientific body that's up to the challenge. The new intergovernmental organisation would exist to marshal evidence on drug resistance and to encourage policy implementation. Working with national governments and international agencies tasked with implementing its recommendations, it would set strict targets to stem the loss of drug potency and speed up the development of new therapies.

An Intergovernmental Panel on Antimicrobial Resistance would be welcome if it allowed for greater coordination of information sharing, surveillance and analysis.

But to whom would this scientific policy recommendation body report? What overarching structure would decide what is to be done and then implement those recommendations?

While distinct problems, one has to presume that, like climate policy, it would require a copy of the IPCC's twin, the conference of parties to the UN Framework Convention on Climate Change (UNFCCC). The IPCC was established in 1988 by the UN Environment Program and the World Meteorological Organization. Four years later, the IPCC played a key role in the creation of its diplomatic corollary, the UNFCCC, a quadrennial space for horse-trading between governments that all but collapsed in 2009 in Copenhagen and which has moved hardly at all since.

We as a species are once again confronted by a difficult issue, with worldwide political and economic implications, and without a global democratic body to address it. And the only option imaginable is a process of technocratic and diplomatic decision-making.

Drug resistance and climate change are hardly the only topics like this. As the IPCC itself proudly declares, the relationship between it and the UNFCCC has become a model for interaction between science and decision makers, and a range of efforts have been mounted in the years since their founding to construct similar assessment and policy processes for other global issues.

In 2012, under the aegis of the UN Environmental Program (UNEP), the Intergovernmental Platform on Biodiversity and Ecosystem Services (IPBES) was established, but in partnership with parties signatory to multiple UN conventions, including those covering biological diversity, endangered species, migratory species, plant genetic resources, and wetlands: an "IPCC for biodiversity." And a similar structure is currently being set up to bring together experts and officials into a subsidiary body of the conference of parties to the UN Convention on Combating Drought and Desertification: an "IPCC for deserts and dustbowls."

For some, even the IPCC/UNFCCC is excessively politicised (read: democratic). Johan Rockstrom, the head of the Stockholm Resilience Centre, and Will Steffen, director of the Australian National University Climate Change Institute, are two of the world's leading climate strategists, and are best known for their development with twenty-six other researchers of the Earth-system concept of "planetary boundaries"—the "safe operating space for humanity" as it relates to climate change, ocean acidification, pollution, ozone depletion, and so on that I discussed a few chapters ago.

Rockstrom and Steffen call for a "global referee" independent of elected governments to ensure humanity does not exceed these boundaries: "Ultimately, there will need to be an institution (or institutions) operating, with authority, above the level of individual countries to ensure that the planetary boundaries are respected. In effect, such an institution, acting on behalf of humanity as a whole."[148]

They suggest the creation of an Earth Atmospheric Trust, "which would treat the atmosphere as a global common property asset managed as a trust for the benefit of current and future generations." But how would the governors of such a trust be picked? Elected by the people of the Earth, or appointed by technocrats?

To be clear: the concern is not over the international aggregation of expertise in a particular topic. Who could oppose such a necessary pooling of knowledge and intellectual resources? Rather, the worry is that we have not properly interrogated this particular IPCC/UNFCCC model nor adequately wrestled with how expertise is imbricated with anti-democratic global governance and its retreat from norms of public accountability, participation and popular decision-making.

Not all those asking questions about the IPCC and UNFCCC's democratic deficit are climate denialists. Indeed, it is precisely those concerned about the ramifications of anthropogenic global warming who should be most worried about the galloping tendency of elites to remove decision-making from direct democratic control and the realm of political contest.

For Harvard science and technology studies researcher Sheila Jasanoff, there are a number of pertinent questions: what is the demarcation line between scientific and political institutions? How do governments construct what she calls "public reason"— those forms of evidence and argument used in making state decisions accountable to citizens? Are these new structures apolitical in service of the general interest, or do they provide unacknowledged protections to particular groups whose interests are at odds with the rest of humanity?[149]

Riffing on this idea, German sociologist Silke Beck and her colleagues ask in a recent paper on the structures of the IPCC and the IPBES that we at least explore "the full range of alternative institutional design options as opposed to implementing a one-size-fits-all model of expertise."[150]

"So far," says Beck, whose research focuses on new forms of environmental and science governance, "no debate has ever taken place about the IPCC's relationship to public policy and to its various global 'publics' or about its normative commitments in terms of accountability, political representation, and legitimacy."

In the last two years, there have been talks among stake-holders on the future of the IPCC, but participants in these closed-door meetings are bound by strict confidentiality agreements, and journalists and researchers have been shut out.

In a parallel fashion, great swathes of legislative topic areas such as monetary policy, trade, intellectual property, fisheries and agricultural subsidies that used to be debated openly in democratic chambers are now drafted, amended and approved in backroom arenas.

It's what sociologist Colin Crouch calls "post democracy": while the pageantry of general elections proceeds, decision-making takes place not in legislative bodies, but in closed-door, treaty-based negotiations between government leaders or diplomats, advised by experts.[151]

In the case of the European Union, the most advanced techno-cratic governance space in the world, we can add to the list of topics outside democratic debate: fiscal policy (that is, all spending decisions) and labour-market regulation, those core policy areas that, apart from defence and policing, perhaps define most what it is to be a state.

Since the advent of the Eurozone crisis, the European institutions have successfully insulated economic decision-making from electorates and shifted it to the junta of experts of the European Commission, the Council of Ministers, the European Central Bank, the European Court of Justice, or even ad hoc self-selecting groups of key players in the European institutional mosaic.

The Eurozone catastrophe was so grave that the EU no longer had time for "political games" or "politicization," as the former European Commission President José Manuel Barroso and Council President Herman Van Rompuy repeatedly stressed. In other words, they no longer had time for democracy.

It's a common sentiment among elites. The incoming commission chief and ex-head of the Eurogroup of nations using the single currency, Luxembourger Jean-Claude Juncker, notoriously said a few year ago: "Monetary policy is a serious issue. We should discuss this in secret, in the Eurogroup," he told a meeting on economic governance organized by the European Movement, not realising the meeting was open to journalists. "I'm ready to be insulted as being insufficiently democratic, but I want to be serious. I am for secret, dark debates."[152]

*

The IPCC/UNFCCC model, the EU, and similar post-democratic structures also operate on the basis of consensus among "stakeholders," rather than majority rule through democratic popular mandate. In other words, policy-making has been globalised, but democracy has not.

Consensus delimits the range of policy options available to those that are amenable to all stakeholders, potentially excluding policy options that may actually solve the given problem if it threatens the interests of a particular stakeholder. The possibility of overruling or even eliminating a stakeholder is precluded by this form of decision-making. The policy window is thus highly circumscribed, and incremental change is favoured over dynamism and innovation. Such policy lethargy is not desirable when it comes to existential threats.

The argument for democracy, then, is not just one of principle. The UNFCCC's post-democratic, consensus-based structure is one of the reasons why climate negotiations are perennially stalled.

And so it would be with a comparable governance model for drug resistance. Farrar and Woolhouse explain that such a strategy is necessary because "the scientific and business worlds need incentives and a better regulatory environment to develop new drugs and approaches."

The pharmaceutical companies are thus considered stake-holders to be welcomed at the table, operators that need to be incentivised to change their ways rather than the key structural obstacle to be overcome. Such incentives include tax credits or grants for priority antibiotic development, "transferable priority review vouchers" that expedite regulatory review for another product of the company's choosing, advance-purchase commitments, and patent-life extensions.

The concept of advance market commitments—in essence, when a government guarantees a market for a successfully developed medicine—is promoted by the World Bank and free-market think-tanks like the Brookings Institution as a solution that fills the gap left by market failures while leaving capital's profits unchallenged.

The most elementary and cheapest solution would be the socialisation of the pharmaceutical sector, permitting the democratic redirection of revenues from profitable therapies to subsidize R&D in unprofitable areas. Prior to privatisation across the West, this cross-subsidisation model permitted postal, rail, bus and telecommunications services to be provided to remote regions, as revenues from the urban centres balanced things out in the interest of universal service.

But such a simple model is not merely off the table because it is politically unrealistic. It is off the table because the very structure of consensus-based intergovernmental and stakeholder decision-making does not allow such solutions to even be raised.

In a clarifying recent paper on the growing preference in some quarters for what he terms "environmental authoritarianism," science and technology policy researcher Andy Stirling writes that "democracy is increasingly seen as a 'failure,' a 'luxury,' or even 'an enemy of nature'... So, knowledge itself is increasingly imprinted by the age-old preoccupations of incumbent power with rhetorics of control. It seems there is no alternative but compliance — or irrational denial and existential doom."[153]

On the contrary, Stirling argues, democratic struggle is the principal means by which sustainability is shaped in the first place—and we should view antibiotics as a precious resource to be carefully shepherded and sustained. "[C]oncentrated power and fallacies of control are more problems than solutions ... among the greatest obstacles to [progressive social transformation], are ideologies of technocratic transition."

A couple of thought experiments to underscore the point: first, French economist Thomas Piketty recently proposed a confiscatory global wealth tax as a solution to capitalism's inherent tendency toward ever-greater inequality. It has to be global, he rightly says, in order to avoid inter-state competition to deliver the lowest tax rates.

But imagine if this policy were taken seriously for implementation. How could such a tax be imposed by any agency other than an elected, global government with a strong mandate to do so? A model based on the UNFCCC or EU structures would end up mired in years or decades of fruitless discussion, at best resulting in a highly watered-down version that all stakeholders could agree to—much like the dismal, foundering effort to introduce a Tobin Tax across Europe.

A second thought experiment: if we discovered tomorrow that a large near-Earth asteroid were on a course for the planet and was due to obliterate human civilisation in five years' time, which would be your favoured mechanism of developing a planetary defence system and mounting a mission to divert it?

A global, democratically elected government that could within weeks pick the best plan after receiving advice from experts and then rapidly direct resources to where efforts would be most efficient and likely to succeed?

Or a series of multilateral stakeholder talks debating for most of those five years who would bear the bulk of the cost (if you're familiar with the "climate finance" debate, try "asteroid finance"); which country would get the most jobs from the project; which companies would win the contracts; how to share data, technology and best practices; and which city would get to host the project secretariat?

*

About fifteen years ago, the global justice movement mounted a critique of this kind of extra-democratic decision-making, focusing on its incarnation in international institutions like the WTO, the World Bank, the IMF and the G8, and in "investor rights" chapters and investor-to-state dispute settlement clauses in trade agreements that permit democratically approved legislation and regulations to be overturned by closed-door, unelected trade tribunals.

Similarly, the struggle today against EU-imposed austerity across southern Europe—often led by veterans of those millennial street battles—also involves a critique of the steady removal of ever larger sections of fiscal policy from the realm of democratic control.

But for the most part, this critique of post-democracy has amounted to little more than a demand for a return of national sovereignty. Globalisation is neoliberal and undemocratic; therefore, we propose the small and local. European integration is austerian and technocratic; therefore, we propose a break-up of the EU.

Conversely, the recognition that existential threats such as drug resistance and climate change must be confronted at the global level often causes well-meaning, pragmatic people to embrace the creation of international, but post-democratic structures.

Yet there is a third option that is both better suited to the task and intrinsically preferable to the status quo: genuine transnational democracy, both at the continental and global level. This means an abandonment of polite but undemocratic stakeholder negotiation between bureaucrats, diplomats and their experts, and the welcome return of robust ideological antagonism, of majority rule, and messy clashes of radically different ideas and programs, of what Stirling calls "open, unruly political struggle"—of democracy.

Existential threats are not just scientific, medical or environmental problems. They are also social, political and economic problems, and that is why democratic struggle is the solution that suits them best.

What precisely this could look like is beyond the scope of this book. Perhaps a UN Parliament from which a global prime minister and cabinet were drawn, with similar models in Europe (meaning a dissolution of the unelected commission and indirectly elected council) and on other continents. The exact contours are not for me to describe anyway: if global governance is to be democratic, then by definition it has to be fought for and built by grassroots democratic movements. It cannot be an elite inspiration or construction.

But it is long past time that we set aside the idea that global government is a utopian—or dystopian—fantasy. Or critique it from a small-is-beautiful standpoint little different to the anti-globalisation perspective of nationalists and New World Order conspiracists. *It's already happening*, and we do need it desperately to deal with the global scale of problems we now face. Global government is here. We need to make it democratic.

Democracy is the Enlightenment sibling of science. It is no barrier to solving problems like antibiotic resistance and climate change. Rather, it is, as it has always been, humanity's best hope.

11

There Is No 'Metabolic Rift'

Most of my best-selling and award-winning targets so far—the likes of Naomi Klein, Bill McKibben, Annie Leonard and Ronald Wright, and their more radically primitivist but no less best-selling or award-winning confederates, Derrick Jensen, James Howard Kunstler and (Booker-prize-longlisted!) Paul Kingsnorth—are essentially mere popularisers of the anti-growth, anti-progress *Weltanschauung*. Alongside their raft of bestsellers and documentaries, there is a much larger, ahem, ecosystem, of NGO working papers, think-tank reports, academic journals, symposia, formal declarations—a sizeable and growing intellectual current—that backs degrowth economics.

The *décroissance* or degrowth movement is far from fringe. Largely European-based, it has its roots in the aforementioned Club of Rome's 1972 *Limits to Growth* report, and has been a key tenet of the field of ecological economics since its establishment in the 1990s, but properly made its mark on public discourse in the late 2000s in Europe around the ideas of Paris-Sud 11 University economist Serge Latouche, and the first 'degrowth conference' in Paris in 2008. The following year saw the publication of *Prosperity Without Growth*, released as a report by Sustainable Development Commission, the UK Labour government's short-lived[154] advisory group on, as its name suggests, sustainable development. Authored by the economics commissioner of the body, University of Surrey ecological economist Tim Jackson, the report was later published as a book that has been translated into 15 languages and was described by *Le Monde* as "one of the most outstanding pieces of environmental economics literature in recent years." Across the Atlantic,

Lester Brown's venerable Worldwatch Institute published its own, similar document in 2012, *The Path to Degrowth in Overdeveloped Economies*[155]. In 2014, Routledge published a thick reader on the subject, *Degrowth: A Vocabulary for a New Era*[156], with contributions from a range of academics in the young discipline. Packed international degrowth conferences have been held in Paris, Barcelona, Montreal and Venice, with dozens more local workshops and seminars mounted further afield. Naomi Klein was keynote speaker at Degrowth 2014 in Leipzig. The Catalonian capital is also host to Research & Degrowth, an academic institute dedicated to "research and actions to consume less and share more."

Just as with the divergences between Klein and Kingsnorth, there is considerable disagreement, if not more so, amongst the academics and wonks, as they feel the need to add some theoretical and policy substance on the otherwise abstract slogan of calling for an end to growth.

The most immediate bifurcation, paralleling the debate of whether we need to dial back to pre-war, pre-industrial or pre-agricultural times, is between the advocates of a downscaling of developed economies while letting developing countries to advance to a 'Western standard of living'[157]; and those that say that even developing countries need to scale back.

Latouche is unapologetic. Guatemala, Somalia and Congo-Brazzaville in his mind are also too advanced. "Degrowth must apply to the South as much as to the North if there is to be any chance to stop Southern societies from rushing up the blind alley of growth economics. Where there is still time, they should aim not for development but for disentanglement ... If the South is to attempt to create non-growth societies, it must rethink and re-localise."

"Insisting on growth in the South, as though it were the only way out of the misery that growth created, can only lead to further westernisation," he continues, playing the anti-colonialist card. "Development proposals are often born of genuine goodwill – we want to build schools and health clinics, set up water distribution systems, restore self-sufficiency in food – but they all share the ethnocentrism bound up with the idea of development."[158]

Clearly, there is a great deal to criticise about neoliberal development models and the resurrection of the White Man's Burden in NGO drag, but really, do visit Sierra Leone or Liberia or Guinea and ask locals whether the construction of public health infrastructure and development of domestic human resources sufficient to be able to identify, track and isolate those suspected of being infected with Ebola—the bare minimum necessary to begin to contain the 2014 outbreak—would be neo-colonial *or a miracle*.

Needless to say, there are other degrowth activists who say this goes too far. Herman Daly, the former World Bank economist and *Adbusters* magazine's Man of the Year 2008, is the godfather of the modern concept of the 'steady-state economy', wherein replacement investment would be carried out, but no new net investment. In his words: "An economy with constant stocks of people and artefacts, maintained at some desired, sufficient levels by low rates of maintenance 'throughput' – that, is, by the lowest feasible flows of matter and energy from the first stage of production to the last stage of consumption."[159]

And still others place themselves somewhere in the middle. Left-wing writer John Bellamy Foster, suggests a rough shrinking of the rich economies "by as much as a third," while the rest of the world catches up to that level.[160] And once that has happened, one aims to maintain a steady-state economy at that level in perpetuity. (Foster is also the editor of the one-time Maoist *Monthly Review* journal that emerged from the US New Left of the 1960s. The modish, western Maoism of this time dismissed the working classes of "the imperialist centre" as a pampered "labour aristocracy" that luxuriated in "superprofits" extracted from the Third World. These *soixante-huitards*, particularly in France and the US, looked instead to the peasant toilers of the global periphery as the new, true revolutionary class, and enthused about the primitivist egalitarianism of China's murderous Cultural Revolution, with its *Up to the Mountains and Down to the Countryside Movement*, in which allegedly privileged young intellectuals were sent to remote villages to unlearn their bourgeois ways and be re-educated through manual agrarian labour. When the grand folly of Maoism fell out of fashion, it was thus not much of an ideological leap for the intellectual current around *Monthly Review* to take up the newly emerging and similarly ascetic ideology of environmentalism.)

The global economic crisis of 2008 has presented something of an awkward problem for some degrowthists, as what they were calling for actually began to happen, and it doesn't look pretty, particularly in the US and EU. Both Tim Jackson and Joan Martinez-Alier, of the Institute of Environmental Science and Technology (ICTA) at the Autonomous University of Barcelona, both leading degrowth theorists, have pulled back and said that, at least for now, for a short period a Green New Deal-style stimulus package would be welcome. But they recognise the contradiction with this moderation and their fundamental critique of growth. As Jackson admits: "the default assumption of even the 'greenest' Keynesian stimulus is to return the economy to a condition of continuing consumption growth. Since this condition is unsustainable, it is difficult to escape the conclusion that in the longer term something more is needed."[161]

What leaps out immediately, regardless of which degree of degrowthist asceticism is preferred, is that all positions seem to be pulled out of thin air, without any empirical grounding. Why a third rather than a half? What is the 'appropriate' level of development? If developing countries can grow to 'our' level, what is so special about the Western level of development [sic] in 2014? Why not, say, a Norwegian standard of living circa 1983? Or if even Ethiopia is over-developed, is Vanuatu over-developed too?

Another question that might be asked: once we have degrown to this 'appropriate' level, whatever level that might be, and we are now just chugging along neither expanding nor shrinking (a remarkable, trapeze-artist-like feat of perfect balance, by the way), what happens when someone invents a brilliant new technology? Remember that we have decided that this, right here, this amount of 'stuff' is all that we need, and no more. Anything new by definition is unnecessary. The logical conclusion of degrowth is ineluctable: we must remain techno-logically, scientifically, medically frozen. All new innovation linked to the material world must be relinquished. The world of *ideas* may advance of course, as they are immaterial, but woe betide anyone who tried to develop those new ideas into new machines.

The immediate response to this would be: but some innova-tions replace old technologies without any additional material cost.

This is true abstractly. But some don't. And most are a bit of replacement of existing processes, but made more efficient, while offering up solutions for brand new, heretofore undiscovered needs. And for most, it's a bit of both. But how is any of this additionality measured anyway?

Other debates include whether degrowthists need to actively work toward winning society to a degrowth perspective or, as the Worldwatch Institute's Erik Assadourian, suggests, as a result of what he believes to be the finitude of the Earth's resources, they just have to sit back and wait for Peak Oil or Peak Water or Peak Phosphorus or just Peak Everything to do all the heavy lifting for us, with degrowth the inevitable result of hitting such walls. Degrowth is inevitable, whether we want it or not, he says.

A more interesting discussion, at least for leftists, is over whether degrowth is achievable under capitalism, or whether degrowth requires capitalism's overthrow.

Given the popularity of the critique of economic growth even at the top of society by such figures as Prince Charles and the Club of Rome, clearly there are many who believe we can indeed keep the free market and a steady-state economy. We can have this much capitalism, but no more. A great many environmentalists critique economic growth but only a tiny number are anticapitalist, preferring instead a sort of small-scale steady-state capitalism, a non-growing, Proudhonian, localist paradise of shopkeepers and farmers. Latouche, for his part, argues that eco-compatible capitalism is conceivable with high levels of regulation to bring down the 'ecological footprint'.

Foster, a Marxist of a certain stripe, argues that one cannot simply posit degrowth as a goal outside of understanding that growth under capitalism is a process of endless accumulation of capital. Degrowth must be recast as *deaccumulation*.

"[We] must aim not merely for degrowth in the abstract but more concretely for deaccumulation – in the sense of a transition away from a system geared to the accumulation of capital without end," he writes.

This is because, even as small-scale capitalists, if the owners of production do not reinvest their profits—realized via the extraction of surplus value from the unpaid labour of their workers—in ways that produce *still more capital*, they are no longer capitalists; they will go bankrupt. "Accumulate! Accumulate! That is the Moses and the prophets," as Marx wrote in the first volume of *Capital*.

No mega-corporations emerged fully formed from the forehead of Zeus; they all started off somewhere small at some point in the past—at one point somewhere, they too were small-scale capitalists of degrowthist fantasy.

So it is true that you cannot have capitalism without growth, but this is not an argument against growth.

Once again we see that despite the likes of Klein explicitly saying that they are anti-capitalist, that we have to do away with capitalism, they appear to have never thought to educate themselves on the 150 years of heterodox economics that has sought to critically explore and explain the problems of the capitalist system. It is a moral rage against the free market, not a rigorous, evidence-based consideration of *why* it does what it does. It's enough that it's big and mean. Boo to big things.

To be fair, there are many mainstream Green groups whose leaderships do not accept the degrowth thesis, saying that with the right regulation and incentives, the global can achieve green growth via a dematerialisation of production, using fewer and fewer resources per unit produced—a process known as 'decoupling'. It is useful to consider the debate at the top of the green movement over the question of growth.

And it is a bitter debate. A debate between two big bears of the movement on degrowth at Zeppelin University in Friedrichshafen, Germany, in March, 2014—Ralph Fücks, the president of the German Green Party's research institute, the Heinrich Boll Stiftung, and André Reichel, a researcher with the European Centre for Sustainability Research the university and leading degrowth activist—descended into name-calling and heated attacks from members of the public. Despite both speakers being members of the party, Fueks's green growth concept was ridiculed as "a neoliberal technocratic fantasy," while Reichel's call for economic contraction was vilified as "an authoritarian socialist vision."

It is far too easy to get caught up in abstract declarations and moralistic or aesthetic arguments about scale, say Fücks and his co-thinkers, who emphasise instead the need to pay attention to the numbers. And when we do, we find that indeed, there is substantial evidence for decoupling, for secular declines in resource and energy intensity everywhere you look.

The free market is all about seeking productivity increases and efficiency gains. Labour and raw material are a cost to the capitalist. So in the chase for profits, ever greater efficiency improvements allow him to shrink input costs. Through techno-logical innovation, logistical refinement, production-process reorganisation and communications improvements, material throughput condenses. This is something that even degrowth researcher Tim Jackson concedes. Indeed, in his major work on the subject, *Prosperity Without Growth*, he offers graph after graph demonstrating widespread declines in energy, carbon and metal ore intensity as GDP increases. Energy and material intensity has declined significantly over the last three decades, across the West in particular, declining three times faster in the wealthy nations over the previous quarter century than in non-OECD countries.[162] Similar figures show a decline in material intensity, although not as sharp as that for energy. Nevertheless, taken together, Jackson agrees, all this demonstrates "clear evidence" of decoupling.

With the right policy mix, including a massive emphasis on innovation, strong push for international technology transfer to developing nations, radical changes to patterns of recycling waste—all of which would be very difficult given our planet's Westphalian hangover and current lack of democratic global governance structures—but certainly not *impossible* to achieve in principle.

But Jackson stresses the importance of distinguishing between *relative* and *absolute* decoupling. And it is crucial here to under-stand what comes next: even as material throughput has declined per unit of production—that is, we use steadily less raw material and energy *per widget produced*—overall, production has increased to such an extent that in absolute terms, more raw materials and energy are consumed.

This is the phenomenon of the Jevons Paradox, or rebound effect, first discovered by English economist William Stanley Jevons in 1865: that with technological advance, resource efficiency gains tend overall to increase rather than decrease consumption of that resource. So, as Jackson notes, while global carbon intensity has fallen over the last three decades from roughly one kilogram per dollar of economic activity to around 770 grams per dollar, *overall*, global CO2 emissions due to energy use have climbed by 40 percent since 1990.

¿Que? Say what? How is this possible? It seems counterintuitive. Shouldn't greater efficiency in the use of a factor of production—whether it be energy or raw materials (or labour)—result in a reduction in demand for that factor? Turns out nope, the opposite happens.

When you think about it though, it makes perfect sense. Efficiency improvements tend to reduce the *price* of a factor of production and increase the *supply* of that factor, which results in companies and consumers finding new uses for that factor. This process (substitution) happens automatically, in an unplanned fashion, a direct product of the price mechanism within capitalism.

What this means is that energy efficiency regulations, while excellent public policy that improves the average standard of living (accepting for aforementioned reasons that 'average standard of living' is a problematic concept), it is a real bugger for climate policy, as it in effect says that increasing energy efficiency in production processes actually *increases* carbon emissions.

A corollary to this is that while efficiency should still be forcefully pursued, any emissions reduction schemes that depend in large part for their success upon assumptions of massive improvements in energy efficiency simply won't work. Instead, the policy emphasis should be on technological innovation, enabling a wholesale shift to carbon-neutral, or relatively carbon neutral electricity production. Once that is achieved, any expansion in absolute energy consumption—the Jevons Paradox—ceases to be a problem as far as global warming is concerned.

Some pro-growth advocates, like Roger Pielke Jr., make precisely this argument: that because growing economies tend to become more resource efficient and efficiency allows us to decouple emissions from growth, the best way to reduce carbon emissions is actually to grow the economy. Alongside this, richer economies can better afford the clean technologies that reduce emissions. Indeed, the longer our governments delay appropriate action on emissions reductions, the more expensive the infrastructural transition to a low-carbon economy will be. Similarly on the adaptation front, the wealthier an economy is—particularly in the developing world where some of the most severe and earliest climate effects are being felt—the more able it will be to afford the cost of adaptive measures such as building flood defences or commissioning new buildings with a climate resilient design. It has been noted many times how droughts in east Africa threaten the lives of millions while those in Australia kill no one. The difference of course is a mystery to no one: the level of development.

With the right global policy mix[163], growing the economy to achieve sufficient energy and carbon decoupling could in principle work (although whether it could happen in time to avert catastrophic or runaway climate change is unanswerable), in order to get us over the top of the infrastructure and technology required for a zero-carbon economy (remember that for genuine *renewable* electricity production, the rebound effect ceases to matter[164]—so long as it's renewable, we can use as much as we want without negative impact on the climate). But outside of a 'zero-waste' production system—which, like a perpetual motion machine, is physically impossible—this doesn't work for *material inputs*—the second aspect of the anthropogenic biocrisis. The Jevons Paradox shows that efficiency gains in manufacturing (or other material production processes) only mean an inexorably greater use of material resources. More investigation into the nature and dynamics of such rebound, and whether it changes from industry to industry would be helpful. Nevertheless, Jevons appears to be an insurmountable problem under capitalism.

However, we can hypothesise that the Jevons Paradox should disappear under the economic planning and absence of the profit motive that would exist under some form of socialism. Instead of production and consumption decisions being led by the blind forces of price in the marketplace, they are consciously, *democratically* decided. One of the fundamental axioms of the socialist worldview is that humans can come up with a superior distribution of goods and services than the market can. This axiom—of which pseudo-anti-capitalists like Naomi Klein, just like neoliberal ideologues, are deeply suspicious—is born of a deeply humanist confidence in the capacity of our species to improve upon 'natural' processes, to repeatedly breach the barriers placed in our way by the rest of nature.

Instead of next investment or production decision being driven blindly by profit seeking, or consumer purchase made constrained by the need to reduce expenditure, all economic actions occur as the result of rational decision-making on the basis of maximum utility to society. Because this all this is a conscious, planned process and we are no longer beholden to the drive for profit, *we would now have the possibility to wait, to hold off for a while until we have sufficient technological innovation to move forward in a way that does not damage the environment in a way that delimits the optimum living conditions for humans.*

We can collectively say: Well, now that we have this new efficiency in the production of this commodity, what shall we do with the savings? Shall we increase production? Shall we reduce material use? Shall we increase the overall amount of leisure time available to the labour force?

Capitalism is a problem because in the face of environmental spoilage, it must proceed regardless (not because of growth per se!). Any new innovation permitting efficiency gains will be invested in the optimum way to produce still more capital, even at the expense of environmental despoilment. This is not to say that the capitalist is evil. He is not. He has no choice. Indeed, even if he is environmentally minded, he must still make that choice, or go bankrupt. As Foster writes, and here he is correct, the constant drive to accumulate capital "impos[es] the needs of capital on nature, regardless of the consequences to natural systems." Democratic economic planning though gives us breathing room. True, we may in principle at some point in the future have to pause some production expansions here or there, for a period. But this is a very different thing from saying there is an upper limit.

Even better, because socialism would permit us to direct investment—including investments in research and development—not merely toward what is profitable, but toward what is most useful, there is every likelihood that growth *may actually advance faster under socialism* than under capitalism, because more research funding can directed to technologies ensuring we do not damage the environment.[165]

*

The Forced Marriage of Marx and Malthus

If we are going to talk about capitalism vs the climate, as the subtitle of Klein's book urges us to do, then we do rather have to consider the pair of eight-ton elephants in the room: those states in Eastern Europe and Asia that used to be known as 'actually existing socialism' before it actually stopped existing; and the hardly much better environmental record of Western social democratic governments in office.

By the time of the Tiananmen Square massacre and the fall of the Berlin Wall, only a tiny number of people on the left in the West considered Moscow or Beijing to have ever been representative of socialism. (Indeed, from my point of view, to have defended the Soviet Union relieves you of the honour of being able to count yourself amongst the left *tout court*.) Following the post-1991 revelations of the true levels of Soviet ecological destruction, for the majority of socialists who had never supported Stalinism to begin with, it was enough to say that just as the regime had cared little for human welfare, so it cared little for the environment. The second followed from the first.

I adhere to this view. It may seem a simple get-out to say that the Soviet Union wasn't truly socialist, but then don't we in effect perform the same manoeuvre when we permit ourselves to be democrats while abhorring the wrong-doings of the early democrats such as Cromwell, Washington and Robespierre? It was not democracy that caused the reconquest of Ireland, the Reign of Terror, or the maintenance of slavery, but its insufficient application. Just so with the still noble idea of one day replacing the blind greed of market competition with cooperation, compassion and planning.

A handful of leftist academics, however, felt pressure to go further still, to make sure there could be no confusion that clear blue ideological water existed between themselves and the USSR's environmental destruction (and latterly, that of the People's Republic of China). To achieve this distancing, they declared that alongside the Gulag, the purges, show trials, and the architectures of famine and unfreedom, we can also add *productivism*—the term used by critics of growth to describe the beliefs of its proponents—to the list of crimes committed by Stalin and Mao. Some argued that the original sin of productivism—sometimes called prometheanism to also encompass a support for scientific, technological and medical advance—had been committed by Marx. Ted Benton, in a seminal such critique, agreed that the old fellow and Engels were guilty of a "'productivist' 'promethean' view of history." This all repeated the long-standing anti-socialist critique of Green Parties. Jonathan Porritt, a founder of what is now the Green Party of England and Wales and ex-director of Friends of the Earth, back in 1984 described socialism and capitalism as Tweedledum and Tweedledee as far as he was concerned, since both "are dedicated to industrial growth, to the expansion of the means of production, to a materialist ethic as the best means of meeting people's needs, and to unimpeded technological development."[166]

Other writers have tried to marry Marx and Malthus, such as eco-socialist theorist and US Green Party activist Joel Kovel, who argues that it is wasn't Stalinism that caused productivism. Rather, it was productivism that caused Stalinism![167]

Kovel, for all his otherwise genuine commitment to socialism, also expresses something of a nasty contempt for ordinary people at one point, saying he considers it a "naïve faith" to think that freed of capitalism, workers would "unproblematically proceed to rearrange production in an ecologically sane way." (Not only that, but just as Theodor Adorno famously hated jazz, Kovel seems to believe that economic growth causes techno music: "We should not seek to become larger within socialism, but more realized. Bach did not quantitatively expand music, making it louder and more insistent like forms of techno-rock [sic] music that mirror capitalist relations; he rather saw more deeply into its possibilities and realized them. So would it be expected for an ecocentric society, where the ideal of growth as such simply needs to be scrapped." Sorry, Joel, but if I can't have dance music, I don't want to be part of your revolution.) He continues:

> Forged at the moment of industrialization, [socialism's] transformative impulse tended to remain within the terms of the industrialized domination of nature. Thus it continued to manifest the technological optimism of the industrial worldview, and its associated logic of productivism – all of which feed into the mania for growth ... [T]he core of socialist optimism ... [is] its historical mission is to perfect the indus-

trial system and not overcome it.[168]

Meanwhile others such as John Bellamy Foster and Paul Burkett wanted to save Marx from the calumny[169] of the likes of Porritt, and have produced a prodigious volume of books and articles in an attempt to clear his name of such accusations, establish his environmental *bona fides* and declare the great man to in fact be an originator of ecological consciousness, prefiguring the green movement[170] via Marx's notion of the "metabolic rift" — an irreparable breach between man and nature that arrives with the advent of capitalism:

> Capitalism is incapable of regulating its social metabolism with nature in an environmentally sustainable manner. Its very operations violate the laws of restitution and metabolic restoration," write Foster, Brett Clark and Richard York in their 2010 book, *The Ecological Rift*, developing the idea of the metabolic rift further. The constant drive to renew the capital accumulation process intensifies its destructive social metabolism, imposing the needs of capital on nature, regardless of the consequences to natural systems.

Marx himself never actually spoke of a metabolic rift.[171] Rather, Foster draws out the conception from a pair of previously unremarked upon quotes from *Capital* volumes I and III on the decline of soil fertility. The idea, drawing on the ideas of 19th Century agricultural chemist Justus von Liebig, is that the further the city lies away from the countryside, the more difficult it is to cycle human poop and wee back into the soil, restoring the nutrients that have been removed.

But of course, ever-extending networks of production have been with us since the advent of agriculture. This problem of recycling waste is not unique to capitalism or industrial society. Murray Bookchin, an anarchist and founder of social ecology emphasised the anti-civilisational logical conclusion of this argument back in 1974, long before Foster developed his metabolic rift thesis: "I am trying through ecology to heal the wound that was opened by humanity's split with nature thousands of years ago." If such a rift does in fact exist, it would not be capitalism that we need to do away with, but all of civilisation and agriculture too.

And indeed, we have a hint of the primitivist consequences of the metabolic rift proponents when we explore how Foster et al propose to resolve the problem: by an embrace of "meta-industrial labour" of those engaged in naturally "rift-healing" work: "those workers, primarily women, peasants, the indigenous, whose daily work is directed at biological growth and regeneration."[172] Ultimately this noble-savage romanticism is little different from the anti-civilisational writings of Derrick Jensen.

What Foster and friends call the metabolic rift regarding soil fertility[173], elsewhere is referred to as 'peak phosphorus'.[174] This Town Mouse/Country Mouse problem is indeed a serious issue that has been investigated by researchers. Phosphorus, like nitrogen and potassium, is an essential element for living organisms. Without it, life is not possible. It cannot be substituted. It is a component of DNA and RNA, cell membranes and permits cellular energy transport. While historically, much human and animal waste was recycled back into agricultural production, less and less is today, with most phosphorus applied back to soil via processed fertilisers, to the point that the joint Australian-Swedish Global Phosphorus Research Initiative (GPRI) reckons that the world has around 30 to 40 years left of readily available phosphorus supplies.

This, like many environmental problems, is a very serious issue that should not be dismissed as Town Mouse vs Country Mouse flapdoodle. However, now that it has been flagged as a grave concern, we should not treat it as the end of the world, but another technical problem that can hopefully be solved just as other environmental problems have been, from clean air and water legislation in most of the West to the arresting of ozone depletion in the stratosphere[175]. Raising awareness of the problem is useful; doom-mongering is not.

First of all, we do not have accurate estimates of the true extent of accessible phosphate rock. We haven't really looked to see if there aren't sizeable reserves somewhere. So the first task is firming up our estimates. Indeed, in the wake of a raft of episodes of media panic over 'peak phosphorus'—reacting to spikes in the price of the commodity in the late 2000s—mining firms reappraised reserves and expanded their investment in the sector, and as a result, supplies have rapidly grown and prices have dropped back.[176]

Nevertheless, over the longer term, we can't just bank on mining more. Beyond this, we need research into how we can better capture and recycle the vast quantity of phosphorus that is just left to escape, from human and animal waste to the inedible parts of crops such as stalks, stems and cobs, to the mammoth amounts of fertiliser that runs off farms and devastates lakes, rivers and estuaries through eutrophication. According to sustainable-resources researcher and GPRI co-founder Dana Cordell at the University of Technology Sydney, of all the phosphorus that is mined, just a fifth is actually consumed by humans in food, with the most of the rest wasted—leaving great potential for this squandering to be arrested and nutrients returned to soil. Already, a number of cities in Sweden require all new toilets to divert urine for reuse, while a range of jurisdictions are experimenting with the processing of wastewater, and technology that improves the efficiency of wastewater treatment and phosphorus recovery is also under development at Arizona State University under the rubric of their *Sustainable P Initiative*. We may also be able to bioengineer some crops to improve their ability to take up phosphorus from the soil, requiring less application of fertiliser. Researchers Roberto Gaxiola and Charles

Sanchez of the same initiative have already developed this trait in tomatoes and lettuce and are now at work extending it to other crops, including corn, rice, cotton and carrots.

No one part of this is enough of course, but taken together, it all points in the direction of how the phosphorus sustainability problem is going to be solved. Realistically, we are talking about international farm-to-fork-to-faeces governance of phosphorus throughout the entire food production and consumption system, coordinating the agricultural, fertiliser, sanitation and waste sectors. Such 'phosphorus security' will likely require brand new and society-wide composting infrastructure—far beyond little green boxes in the kitchen and backyard that we have gotten used to in the last few years—that simultaneously prevents the dispersion of bioaerosols that contain fungal spores and bacteria that present a hazard to human health.[177] This is no small endeavour. But neither is it an impossible feat, even within capitalism (albeit plainly with a major role, once again, for the public sector and for regulation).

We are so used to the doom-mongering (and there are more than enough genuine environmental problems to understand why the doom-mongering is going on), that we just accept without question the latest peakist warning of cataclysm. Foster's metabolic rift conjecture has been widely accepted on the left, well beyond eco-socialist circles, without question. But few really bothered to look into whether soil science and waste management technologies had developed since the 19th Century. It turns out however that there just isn't any soil-fertility-based metabolic rift, or at least not one that can't be repaired with a bit of good old fashioned innovation and political reorganisation.

12

Small Is Not Beautiful

Leading environmentalist Bill McKibben is perhaps the most well-known latter-day evangelist of *small*. The founder of 350.org, the international green NGO that has the very tightly defined goal of reducing atmospheric CO_2 levels to 350 parts per million, and one of the first to warn of the problem of global warming, published *Eaarth*[178] in 2010, a guide to "making a life on a tough new planet." Written in the wake of the collapse of the 2009 UN climate talks in Copenhagen, it is another pessimistic call for retreat from progress, from growth, for an embrace of the small and modest, declaring that on this new planet, "growth may be the one big habit we finally must break." A former staff writer for the *New Yorker*, he is also a handsome wordsmith:

We're so used to growth that we can't imagine alternatives; at best we embrace the squishy *sustainable*, with its implied claim that we can keep on as before. So here are my candidates for words that may help us think usefully about the future: Durable, Sturdy, Stable, Hardy, Robust. These are squat, solid, stout words. They conjure a world where we no longer grow by leaps and bounds, but where we hunker down, where we dig in. They are words that we associate with maturity, not youth; with steadiness, not flash.

Elsewhere, in a section that almost echoes the anapaestic tetrameter of Dr Seuss (channelling *The Lorax*, perhaps?), McKibben continues:

Most of all, of course, our time has been the time of bigness – the amazing ever-steepening upward curve, where things grew and grew and then grew some more. Economies and road networks and houses, inflating until there were entire subdivisions filled with starter castles for entry-level monarchs. Stomachs and breasts and lips, cars and debts, portions and bonuses. *Can we imagine smaller?*

After a certain point, bigness spells trouble, he says, because useful feedback diminishes as scale expands. But small things breed a kind of stability. He wants more small businesses, small farms, small banks and small power companies. "Better the Fortune 500,000 than the Fortune 500," he wishes, the conservative fever dream of the yeoman and shopkeeper. An earlier book of McKibben's, 2008's *Deep Economy*, also makes the case for degrowth, and localist food and micro-energy production. It seems like there's enough, ahem, growth in the degrowth book market for two books from the same author about the same thing.

Declaring himself both a patriot and a dissenter, he declares the American revolution to be:

the defence of the small against the big ... That is, the Minutemen were, at the outset, defending less the idea of America than the idea of Chelmsford, of small and tight and connected communities. Of small and local economies without the margin to easily afford the various taxes and duties the British imposed. Of the idea that they should be able to figure out their own destiny – which at the time they hashed out in local town meetings each spring.

Sneering at how we once put a man on the moon, McKibben compares such national projects of bigness to those roadside attractions you find across North America, like the world's biggest fibreglass cow, "the world's biggest coffeepot, the world's biggest strawberry, watermelon and artichoke":

Theoretically, we're committed to sending a man to Mars, but I know very few people who either believe we will or care … The project we're now undertaking – maintenance, graceful decline, hunkering down, holding on against the storm – requires a different scale. Instead of continents and vast nations, we need to think about states, about towns, about neighbourhoods, about blocks. Big was dynamic; when the project was growth, we could stand the side effects. But now the side effects of that size – climate change, for instane – are sapping us. We need to scale back, to go to ground.

McKibben, like Klein and the rest, is of course hardly the first to declare bigness to be the root of all evil. They all consciously or unconsciously draw upon the seminal 1973 collection of essays by British economist E.F. Schumacher, *Small is Beautiful: a study of economics as if people mattered*, assessed by the *Times Literary Supplement* as the 17th most influential book published since World War Two[179]. After a career with the British Coal Board, in 1955, Schumacher visited Burma, working as an advisor to the government. While in the country, he had an epiphany that that he described as 'Buddhist economics'—a vision of an economy designed placing harmony, community and ecological values at its centre. Core to the philosophy was small-scale, localist, labour-intensive production, self-reliance; and 'people-centred', community-scale *appropriate technology* (originally termed 'intermediate technology'). What does and doesn't count as appropriate in the latter concept isn't always clear, ranging from pico-hydro electricity generation to bike-powered water pumps to herbalist tinctures, animal-powered transport and cob houses— but also LED flashlights and photovoltaic solar panels—but advocates know it when they see it. Similar to the concept of 'conviviality' from Austrian theologian Ivan Illich, also writing in the 1970s, who argued that advanced technology was the preserve of elite groups with a monopoly on knowledge that "robs peasant societies of their vital skills and know-how."[180] Schumacher founded what would these days be called an NGO, the Intermediate Technology Development Group, which aimed to put into practice his thinking. The group still exists, now operating under the name Practical Action, and is just one of a family of organisations in the UK—including the New Economics Foundation, the Centre for Alternative Technology, and the Soil Association, the British organic farming organisation—that refer to themselves as the Schumacher Circle. Founded in his memory or inspired by his work, the Circle organisations cooperate informally to support each other's work.

A convert to Catholicism, Schumacher recognised the parallels between his philosophy and the culturally traditionalist and agrarian Catholic economic ideology of distributism formulated by G.K. Chesteron and Hilaire Belloc. A simultaneously anti-socialist and anti-capitalist philosophy that was also critical of the Enlightenment, distributism emphasised local culture and opposed mass production, viewing the just social order as being achieved by ensuring that property is spread widely, with a maximal amount of small-holders, artisans and shopkeepers, so that people can earn a living without having to rely on the state (socialism) or a small number of individuals (capitalism). It was a sort of English Catholic micro-Toryism, and indeed distributism influenced continental post-war European Christian Democratic parties, and the youthful US conservative commentators Ross Douthat and Reihan Salam describe their prescription for renovating the GOP for the 21st Century as distributist. Phillip Blond, the conservative political philosopher and counsellor to UK Prime Minister David Cameron pitches an updated version of distributism that he told the Washington Post is a mash-up of Occupy Wall Street and the Tea Party, emphasising neither markets nor government, but what is small and local.[181] "Distributism is very closely related to what we now call environmental and ecological questions," wrote Schumacher's friend and

Catholic writer Christopher Derrick.

*

Blood & Soil

It is fruitful here to consider the person of Jorian Jenks, one of the Soil Association's founders, the editor of their journal, *Mother Earth*, and a self-described "critic of modern economy." A dominant figure in the founding of the organic movement, Jenks was also the British Union of Fascists' agricultural advisor and a close associate of its leader, Oswald Mosley, authorised to carry on the work of Mosley in the event of his arrest. Jenks was inspired by the catastrophist German historian Oswald Spengler's 1918–23 bestseller *The Decline of the West*, in which the proto-Nazi intellectual argued that Western civilisation was experiencing its final season—its wintertime. According to a Graham Macklin, a biographer of Mosley, Jenks also "harboured a congenital dislike and suspicion of science" and sought to "replenish the bond between man and soil."

Rural historian Richard Moore-Coyler describes the catastrophist panic of the interwar right:

Underlying its concerns and anxieties was the belief in a general cultural decline, born of industrialization and urban-ization, which was somehow reducing the nation's vigour and sense of purpose and which, unless countered, would steadily erode the structure of society itself. As the social conse-quences of economic growth, and more especially urban-ization, became self-evident, so the Right looked to the countryside as a renew-able source of vitality which would serve as a 'spiritual' antidote to the perceived dislocation of city life. National regeneration, it was believed, might be achieved by a re-examination of the nation's rural roots; a sort of revival of the agrarian tradition wherein lay the 'true' spiritual strength and cultural and moral virtues of the British people. This rural-nostalgic and usually organicist theme formed a common thread woven into the policies of most ultra-Right groupings of the 1920s and 1930s.[182]

Sound familiar? How alike this all seems to the anti-modernism of the contemporary crunchie left! But hold on to your fork; there's pie! It gets much, much better. Jenks had already been involved with Kinship in Husbandry, a precursor of the Soil Association and national socialist club promoting ruralism and self-sufficiency founded by Nazi sympathizer Rolf Gardiner. Macklin writes that along with Jenks, its members "percolated" the Soil Association, where as editorial secretary he promoted in his words "an anti-modernist philosophy ... the paramountcy of agriculture, the subordination of mechanization to organicism, the localization of economies and the cultivation of a consciousness of ties of blood and soil."

This is of course a direct reference to the Nazis' anti-industrial *Blut und Boden* (blood and soil) doctrine of SS *Obergruppenführer, Reichsbauernführer* (Reich Peasant Leader) and Reichsminister of Food and Agriculture Richard Walther Darré. Blut und Boden embraced a back to the land attitude and called for the re-adoption of rural values and skills. Blood and Soil idealised a 'peasant nobility', with their capacity for loyalty, community, morality and of course their racial purity, who represented the opposite of all that had been lost or ruined by the cosmopolitan, urban, Jewish intellectuals.

Beyond Jenks, the Soil Association Council also enjoyed the company of other far-right figures, including Archibald Ramsay, who was interned during the war for his Nazi sympathies. Indeed until the organisation shifted leftward in the 1960s after Jenks' death, the Soil Association had been a creature of the far right, as concerned with far-right agitation as with ecological concerns.[183]

It is more than disconcerting to anyone familiar with the contemporary green left, New Age eco-mysticist or alternative health scenes to stumble across the following remarkable quote from Nazi botanist Ernst Lehmann, again uncovered by Staudenmeier. Until the final clause of the last sentence, you could be forgiven for thinking it came from a leaflet at a 21st Century local farmers' market (or for that matter, a book by Derrick Jensen or Paul Kingsnorth):

> We recognize that separating humanity from nature, from the whole of life, leads to humankind's own destruction and to the death of nations. Only through a re-integration of humanity into the whole of nature can our people be made stronger. That is the fundamental point of the biological tasks of our age. Humankind alone is no longer the focus of thought, but rather life as a whole ... This striving toward connectedness with the totality of life, with nature itself, a nature into which we are born, this is the deepest meaning and the true essence of National Socialist thought.

Contemporarily, we see the brown ecologists of Umwelt und Aktiv, an environmentalist magazine in contemporary Germany linked to the neo-Nazi National Democratic Party (NPD), and infiltration of the organic movement, by the far right. The state of Rhineland-Palatinate is sufficiently concerned at the trend that it has published a pamphlet for organic farmers on how to combat the phenomenon.

In Italy too, since 2009, we have seen the rise of the Five Star Movement (M5S) of 'anti-politics' comedian Beppe Grillo, who manages to combine opposition to EU austerity politics with localist environmentalism, degrowth economics and anti-immigration rhetoric. Grillo himself has flirted with far-right politicians domestically while in the European Parliament, M5S sits with a grouping that includes the UK Independence Party and the far-right Sweden Democrats.

Meanwhile across the Atlantic, in 1978, John Tanton, the former chairperson of the National Population Committee of the Sierra Club, one of the largest and oldest American environmental groups (and, incidentally, the publisher of Paul Ehrlich's *The Population Bomb*), founded the Federation for American Immigration Reform (FAIR), designated in 2007 as a hate group by the Southern Poverty Law Center. In the 1980s, Ehrlich's wife, Anne, along with a number of other Sierra Club members, urged the organisation to support immigration reduction.[184] In 2004, anti-immigrant candidates almost won a majority of the Sierra Club off the back of an argument that the US standard of living is not sustainable. Similarly, Dave Foreman, a co-founder of the deep ecology movement Earth First! and more latterly the Rewilding Institute, also in the 1980s advocated a reduction of the world's population by denying food aid to Ethiopian famine victims and banning all immigration to the US. Today, along with Earth Policy Institute founder Lester Brown, Foreman is a leader of Apply the Brakes, a California-based ecological and anti-immigration lobby group.[185] Virginia Abernethy, the president of the Carrying Capacity Network—another resource conservation and immigration reduction group—is a self-described white separatist[186]. Environmental philosopher David Skrbina, writing on the Rewilding Institute's website in 2013, explains why green campaigners need to get over the 'immigration taboo':

The remaining piece of the eco-justice puzzle is figuring out how to live on the half of the landscape that we allocate to ourselves. There are two ways to approach this. If we insist on continuing with our present luxurious lifestyle, and its concomitant 25 acre footprint, we can certainly do this. But it means that our 1 billion acres can only support 40 million people—an 87% reduction from current levels. Too draconian?

Too 'radical'? Not really. If we allowed ourselves 100 years to bring the population down, this would require only a 2% annual decrease. This could be easily achieved with a rigorous anti-immigration policy and an aggressive family planning regimen—including, perhaps, government-paid contraceptives, abortions, and sterilizations.[187]

But the argument I am trying to make here is of course not that all contemporary organic farmers are fascists, that all environmentalists are anti-immigrant racists. This is plainly false and to suggest otherwise tips over to green-baiting. The vast majority of eco activists in the 21st Century remain in most other regards highly progressive. You will regularly see the same campaigners at climate camps and immigrant solidarity rallies.

Rather, the point is to recognise the anti-modernist ideological overlap between contemporary green back-to-the-land ideology and *volkisch* agrarian mystique, resulting from common romanticist origins that were deeply antipathetic toward the Enlightenment. Such patterns of green xenophobia keep appearing over and over not due to environmental concern, but specifically because anti-modernism and the logic of limits leads inexorably to population control and immigration restrictions. And while community spirit is certainly a harmless delight, an anti-modernist political emphasis on returning to traditional customs or localist economics in opposition to outside products or influences is by definition exclusionary, no less so than nationalism. It is important for those who quite rightly care deeply about the threat to humanity represented by myriad ecological problems to inoculate themselves against such thinking, to foreswear anti-modernism and the lifeboat politics of limits to growth.

Since the time of the scientific revolution and the political revolutions that accompanied it—most especially the French Revolution—elites whose privileges seemingly daily were being eroded have reacted forcefully both ideologically and physically to this new modern era. The liberal philosopher Isaiah Berlin wrote of a "Counter-Enlightenment" tradition dating back to the 18th Century, a body of ideas that arose in reaction to the rationalism, empiricism and universalism of the Age of Reason. Berlin emphasised the role of the German Romantic movement, which looked to the Medieval period as a simpler, more integrated time. Leftist Israeli historian Zeev Sternhell goes even further than Berlin, describing this intellectual tendency as the 'Anti-Enlightenment'. Berlin was willing to offer the Counter-Enlightenment a tip of the hat for correcting what he felt were the excesses of Enlightenment's tendency toward hubris. Sternhell, a scholar of the rise of fascism and recently thrust into the spotlight when a West Bank settler set off a bomb at his residence and for his comparison of the atmosphere of contemporary Israel to that of 1940s France and his warnings of the collapse of Israeli democracy, gives this intellectual tendency no quarter. He describes the Anti-Enlightenment tradition as a "second modernity," but one that "revolted against rationalism, the autonomy of the individual, and all that unites people: their condition as rational beings with natural rights." For Sternhell, since the 18th Century, the world has been engaged in a ferocious battle between Enlightenment and Anti-Enlightenment modes of thought.

Peter Staudenmaier, German history professor at Marquette University in Milwaukee, who has explored the marriage of nature mysticism and nationalism under the aegis of Prussian anti-Enlightenment irrationalism in the 19th Century, highlights the writings of two key figures, Ernst Moritz Arndt and Wilhelm Heinrich Riehl. The latter is known for his nationalism, but, as Staudenmaier writes:

His remarkable 1815 article *On the Care and Conservation of Forests,* written at the dawn of industrialization in Central

Europe, rails against shortsighted exploitation of woodlands and soil, condemning deforestation and its economic causes. At times he wrote in terms strikingly similar to those of contemporary biocentrism: 'When one sees nature in a necessary connectedness and interrelationship, then all things are equally important—shrub, worm, plant, human, stone, nothing first or last, but all one single unity.'[188]

Staudenmaier finds other striking passages in the work of Arndt's student, Riehl, an 1853 essay, "Field and Forest" which "ended with a call to fight for 'the rights of wilderness.' But even here nationalist pathos set the tone: 'We must save the forest, not only so that our ovens do not become cold in winter, but also so that the pulse of life of the people continues to beat warm and joyfully, so that Germany remains German.'"

We can continue this narrative through the *volkisch* Artamen League of the 1920s, named after the concept of *Artamanen* or 'agriculture man' of blood-and-soil agrarian writer and biologist Willibald Hentschel who argued that his countrymen should retreat from the decadence of urban life and return to a rural idyll, to the aforementioned Oswald Spengler and his fellow ideologue of the Weimar Revolutionary Right, jurist Carl Schmitt, who raged against the "asphalt culture" of modernity and preference for Volkgemeinschaft (folk community) rather than class struggle. Many scholars have articulated the centrality of the apocalyptic, of a loathing of modernity running through all varieties of fascism, and agrarian retreat as solution.

We should be careful not to over-emphasise the German role in the history of Counter-Enlightenment ideas. Berlin also speaks of the Anglo-Irish political theorist and philosophical founder of conservatism, Edmund Burke, "respectful towards church and state and the authority of aristocracies and élites sanctified by history, these doctrines clearly constitute a resistance to attempts at a rational reorganisation of society in the name of universal moral and intellectual ideals," and the "abhorrence of scientific expertise inspired radical protest in the works of William Blake."[189]

Canadian literary theorist Northrop Frye wryly noted that despite Spengler's argument having been disproved multiple times, it remained "one of the great Romantic poems." Frye believed that the declinist meme has been kicking around since the time of Napoleon. The anti-modernist *Weltschmerz* is an inescapable element of modernity: "The decline of the West is as much a part of our mental outlook today as the electron or the dinosaur, and in that sense, we are all Spenglerians."

Yet socialists of the 19th Century such as Friedrich Engels and Karl Marx saw themselves as continuing the Enlightenment tradition, embracing a staunch faith in progress and a conception of history as the steady expansion of human capabilities through rational political economic organization certainly, but also via the emancipatory possibilities of science, technology and medicine, securing ever greater material comforts, banishing irrationality and superstition. Freed of the shackles of capitalism, humanity would stride ever forward out of the realm of necessity and, through greater and greater realisation of our potential, into a cosmos of ever-expanding freedom. Not blind to the hypocrisies of many individual Enlightenment thinkers such as Locke, Smith and Newton, socialists criticised their apologetics and errors of liberal naïveté. Yet their position was not an abandonment of Enlightenment, but a correction.

So how did we get from there to here, where Enlightenment and modernity are cast as villains not only by the various itera-tions of the Counter-Enlightenment Right since the 18th Century, but also by so many figures on the Left? If we are to rehabilitate progress, Enlightenment, modernity, growth and ambition—*prometheanism*—we need to ask where the left took a wrong turn.

And when we do, we cannot avoid the recognition that Theodor Adorno and Max Horkheimer's *Dialectic of*

Enlightenment, completed in 1944 and published in 1947—one of the seminal texts of critical theory and perhaps the central text of the Frankfurt School—is the bridge that links the conservative Counter-Enlightenment anti-modernist reaction to the contemporary growth-fearing, limit-embracing leftist worldview. It would go on to have far-reaching influence on the New Left of the 1960s, and from thence the dominance of post-modernist scholarship in the humanities in the 1980s and 1990s. I am not going to put forward an extensive critique of the ideas of Adorno and like-minded thinkers, as this would simply be a repetition of critiques that have been already developed elsewhere by a variety of far better authors than I. I am simply interested in a sketch of the authors' ideas insofar as this is useful for unpacking the origins of the contemporary degrowth, progress-sceptic, anti-civilisational mood on the green left.

Dialectic of Enlightenment was a product of understandably deep cultural despair toward the end of the Second World War. Adorno and Horkheimer asked themselves what could possibly explain the depth of horror that had been mounted by Nazism. Man had been a wolf to man throughout history, but never on such a vast, industrialised scale. Similar pessimism about the human condition had first set in amidst the unprecedented savagery of the First World War. And indeed the anguish of Adorno and Horkheimer can be read as a further descent into melancholia beyond even the psychic abyss that accompanied the slaughter in the fields of Flanders.

"Enlightenment, understood in the widest sense as the advance of thought, has always aimed at liberating human beings from fear and installing them as masters. Yet the wholly enlightened earth is radiant with triumphant calamity," they wrote, opening the book.

Their conclusion, one of the most thorough-going critiques of modernity yet to have been mounted, was that the horrors of the 20th Century[190] were no aberration. Auschwitz and Hiroshima were the inevitable result of the pursuit of Enlightenment. Tracing this pursuit back to Ancient Greece and Hebrew scripture, they find a repeating pattern throughout history: the hubristic urge toward Enlightenment is unavoidably and precisely a desire for domination of external nature, humans' internal nature, and domination of society.

"Terror and civilisation are inseparable," they concluded. "It is impossible to abolish the terror and retain civilization." In this work, and Adorno's 1951 work *Minima Moralia*, what once was a Marxist critique of capitalism metastasises into a defeatist, fatalist critique of Western civilisation in its entirety. But how is a critique of Enlightenment to be mounted outside of Enlightenment? All possibility of emancipation now seems lost. "That this harshness might be moderated in the future ... seems no more than a dream."

There appears no way out for humanity. In *Minima Moralia*, Adorno wonders whether the only hope for redemption from civilisation's tyrannical urge for dominion is via a kind of tranquil, almost Buddhist, pastoralist relinquishment of want:

Perhaps the true society will grow tired of development and out of freedom, leave possibilities unused, instead of storming under a confused compulsion to the conquest of strange stars. A mankind which no longer knows want will begin to have an inkling of the delusory, futile nature of all the arrangements hitherto made in order to escape want, which used wealth to reproduce want on a larger scale. Enjoyment would be affected, just as its present framework is inseparable from operating, planning, having one's way, subjugating. *Rien faire comme une bête*, lying on water and looking peacefully at the sky, 'being, nothing else, without any further definition and fulfilment,' might take the place of process, act, satisfaction.

How similar this sentiment is to that of the primitivism of Derrick

Jensen! Or for that matter, anarcho-primitivist philosopher John Zerzan's belief that even the invention of symbolic thought or representative language is a practice of domination and form of violence, calling instead for a "non-symbolic consciousness" that he imagines Neanderthals must have had![191]

We find a kindred anti-modernist bent in yet another foundational text of the New Left, critical theorist Herbert Marcuse's *One-Dimensional Man: Studies in the Ideology of Advanced Industrial Society*. The book, published in 1964 and widely described as one of the most important texts of the 20th Century, established for the first time the concept of consumerism as social control. A guru of the 60s street-fighters and counter-culturalists, Marcuse declared class struggle to have been extinguished due to the consumer society's soporific effects, as well as those of the mass media, advertising and managerialism. "The people recognize themselves in their commodities; they find their soul in their automobiles, hi-fi sets, split-level homes, kitchen equipment. The very mechanism which ties the individual to his society has changed, and social control is anchored in the new needs which it has produced."[192] Such false needs breed integration into the capitalist system,

Marcuse extends arguments of Adorno and Horkheimer regarding the Enlightenment to the question of technology. He declares that "technological rationality" and the "logic of domination" that hides within the idea of technological progress impoverishes all aspects of contemporary life. Technological advance only leads to more subjugation:

> Technology, as a mode of production [sic], as the totality of instruments, devices and contrivances which characterize the machine age is thus at the same time a mode of organization and perpetuating (or changing) social relationships, a manifestation of prevalent thought and behaviour patterns, an instrument for control and domination.

We see here in Marcuse multiple parallels with the 1930s anti-technology writings of American historian and civilisation critic Lewis Mumford, with his belief that it was technology—indeed the clock—that caused capitalism, and his concept of the megamachine, vast hierarchical machine-like organisations that use humans as its parts. A critic of the city as well, he argued that the vast Megalopolis would collapse under its own weight into Necropolis, the city of the dead. Mumford, unsurprisingly, is regularly referenced as a visionary by contemporary deep ecologists such as Derrick Jensen and Paul Kingsnorth. Marcuse, like Mumford, also prefigures the neo-luddite positions of anarcho-primitivist philosopher John Zerzan and SDS activist turned advocate of small government Kirkpatrick Sale.

There is also a similar thread of radical anti-prometheanism in the writing of another of the 20th Century's most important Continental philosophers, Martin Heidegger, who in his 1954 essay "The Question Concerning Technology" attacked what he viewed as the monstrous nihilism of modern technological culture, the technological mode of being not merely for its destruction of nature, but for its reduction of all to resource merely waiting for exploitation. All is but an instrument for other ends—without end—obliterating any sense of awe and wonder. The Rhine River ceases to be a river, but instead a source of hydroelectric power. Contrasting modernity with pre-modern, traditional artisanship, Heidegger says that the latter was acceptable because it is not gripped by this instrumentality but rather made use of natural materials in a way that 'brings forth their essence' and that of their environment. In his final text, on the work of German romantic lyric poet Friedrich Hölderlin[193], written just a few days before he died in 1976, he declared: "Because it is necessary to think about whether and if in the age of the technologized uniform world civilization *Heimat* is still possible," referencing the German concept of Heimat—an untranslatable word somewhere between 'home' and 'homeland'.

Cosmopolitan modernity[194] cast against nature, regional identity and *volk* community. We might recognise here an affinity with the petty bourgeois localist, small-is-beautiful mindset, but also *Blut und Boden*. Are we at all surprised about Heidegger's scandalous membership in the NSDAP anymore?

And, influenced by this thought, the New Left in America, from its birth in the early 1960s, turned away from the expansive ambition and productivism of earlier social democratic and Communist left, and toward this fear of the large, the modern, the technological. The 1962 Port Huron Statement of the Students for a Democratic Society (SDS), the 25,700-word founding document of the American New Left, was a salmagundi of concerns over "supertechnology," overpopulation , "uncontrolled housing," and the "problems of the city." For all its anti-racism and solidarity with the "worldwide outbreak of revolution against colonialism and imperialism," it is fundamentally a conservative, anti-urbanist document.

And this general attitude established by these thinkers has continued to frame so much of the Left ever since, from what we think of as 'The Sixties', or more specifically the events of '1968'—a year that continued for some time after its calendar ended, from the street-fighters of Paris in May to the riots of Chicago to the 'hot autumn' of industrial Italy the following year—up to the anti-globalisation movement of my formative years and the return of street battles from Seattle to Genoa, and the horizontalism of Occupy in 2011. Meanwhile, the anti-universalism, anti-positivism, relativism and science-scepticism of the postmodern academy, with its slander of Enlightenment as imperialist and Eurocentric[195], opposition to 'grand narratives', and imprisoning of the word *truth* in scare quotes, continues the philosophical project of Heidegger, Adorno and friends, a project of modesty and regress. Two generations of student activists now, or at least those studying the humanities, have been trained in such anti-utopian, anti-rationalist thinking.

I do not want to dismiss this left by any means. It is my left, the one I grew up in.[196] And it has achieved a great many feats advancing the general welfare: It is this left that won so many civil rights battles in America, renewed the struggle for women's liberation, launched the gay rights movement and placed environmental problems firmly at the centre of the international conversation. But just as sections of the earlier left made the mistake of not seeing the Soviet Union for the macabre wretch that it was, this now quite old, fifty-something New Left has made its own ideological mistakes. And the primary mistake we have made is the turn away from Promethean ambition and embrace of this basket of conservative, romantic small-is-beautiful ideas.

Chinese-American geographer Yi-Fu Tuan wrote in 1974 that back-to-the-land arguments appear to have been with us since the rise of the town, finding instances of such anti-urban attitudes in urban settings, of the desire of the civilised to escape civilisation, in ancient Greek and Roman writings, and even in the Epic of Gilgamesh from Mesopotamia—humanity's first great piece of literature. This recurring pattern of a wistful, sentimental appreciation of nature and lamentation of a lost Eden arises from a certain level of city-dwelling privilege forgetful of the tribulations of rural life and ever-present menace that is the wilderness. It takes a certain kind of forgetfulness to be able to romanticise the hard-knock life of the peasant. The peasant would trade places with the gentleman horticulturalist—or, more latterly, the Stoke Newington subscriber to *Modern Farmer* magazine—any day.

We learn in school that the Enlightenment was an intellectual rupture that happened in Europe and America in the 17th Century, but it is better described as a recurring impulse in humanity toward reason over received authority and tradition than a definite period in history. Bertrand Russell in his history of Western philosophy pointed to its origins in antiquity. Others, stressing this universal urge rather than Enlightenment's specifically European origins, have pointed to thinkers and movements pushing for a similar basket of ideas as far afield as ancient Persia, Judea, Arabia, Tang dynasty China, and early modern India.

If the drive toward Enlightenment has always been with us and, following Yi-Fu Tuan's argument that reactions against urbanity have always been with us—that anti-modernity in effect existed long before the modern period—we can see how the argument against economic growth and allied anti-Promethean ideas is ultimately not in any way a *novel* phenomenon, but just another iteration of ancient anti-Enlightenment reaction. It is the latest episode in what appears to be an eternal battle between the proponents of human progress and its conservative opponents. It is a battle between those who recognise that within our species while there is the potential both for greatness and for wickedness, overall, we are ascending, and those who focus only on our wickedness and believe us to have fallen long ago.[197]

Along these optimistic lines, Steven Pinker, the liberal evolutionary psychologist, has written one of the most important, if flawed, works of historical research ever published, 2011's *The Better Angels of Our Nature*. It is in essence an 832-page review article, drawing upon the work of many other researchers across many fields in which he patiently, carefully expounds how, contrary to the popular belief that society is increasingly violent, robust statistical evidence and quantitative historical investigation shows how the human story is one of steadily *decreasing* violence. Drawing on scholars' work tracking court and county records in England, he writes how they have found that homicide rates have plunged from 110 homicides per 100,000 people per year in Oxford in the 14th Century to fewer than one homicide per 100,000 in London in the middle of the 20th Century. Research covering Italy, Germany, the Netherlands,

Scandinavia and Switzerland, reports similar trends. Expanding the geography under consideration, Pinker shows how violent deaths of all description have plummeted, from roughly 500 per 100,000 people per year in prestate societies to roughly 50 in the Middle Ages, to six to eight today globally, and fewer than one in most of Europe. Importantly, he notes that this evidence says nothing about how smooth the path downward has been or that this pattern will continue.

This of course is fantastic news. If human civilisation can produce ever greater cooperation and empathy and a colossal reduction in violence, it follows that there is nothing in principle—nothing inherent to human civilisation—to prevent us from cooperating to resolve the climate crisis and parallel environmental problems. It is time to abandon the pessimism, crippled ambition and human-hating of Heidegger, Adorno, the Counter-Enlightenment, the postmodernist academy and the primitivist left to return to the essence of socialist humanism: a celebration of our species' proven capacity for moral and material improvement.

Against the degrowth advocates, critics of progress and the Enlightenment, those who favour an improvement in the human condition must rehabilitate Prometheanism—the idea that there are no limits other than the laws of physics to how we can re-engineer ourselves and the world around us. We, alone in nature, possess the ability to condition that which conditions us, giving us near infinite malleability. There is no fixed human nature; we are not 'dehumanised' (that is, some human essence removed) by technology or modernity or civilisation. This Promethean conception is not alien to the left; indeed, to give our plasticity full rein, it requires a breaking with the current mode of production.

"When the limited bourgeois form is stripped away," wrote Marx in the *Grundrisse*, what is wealth other than the universality of individual needs, capacities, pleasures, productive forces created through universal exchange? Full development of human mastery over the forces of nature, those of so-called nature as well as of humanity's own nature? The absolute working-out of his creative potentialities, with no presupposition other than the previous historic development, which makes this totality of development, i.e. the development of all human powers as such the end in itself, not as measured on a *predetermined yardstick?* Where he does not reproduce himself in one specificity, but produces his totality? Strives not to remain something he has become, but is in the *absolute movement of becoming?*[198]

Such an epoch, this "true realm of freedom," depends upon the wealth of society constantly expanding, permitting over time for us all to work less and less, or for the line between work and leisure to begin to blur. This is why the socialist must defend economic growth, productivism, Prometheanism. The long-standing promise of socialism was not that we'd have the same stuff as under capitalism, but shared out equally; rather it was that through equality, we could release the forces of production from the fetters place upon them by capitalism. We would have so much *more stuff* that under capitalism.

In *All that is Solid Melts into Air*, the late American philosopher and socialist Marshall Berman's classic celebration of fizzing, thrusting, transfiguring, luminous, *contradictory* modernity (and paean to his beloved New York), he describes the modern condition perfectly: "To be modern is to find ourselves in an environment that promises us adventure, power, joy, growth, transformation of ourselves and the world – and, at the same time, that threatens to destroy everything we have, everything we know, everything we are." Berman cheers "human adventure, progress, faith in the future, all the heroic ideals of the age into which I was born" and condemns the "widespread and often desperate fear of the freedom that modernity opens up for every individual."

It's often forgotten, but Marx and Engels in *The Communist Manifesto* express great admiration for the best of what capitalist dynamism had wrought by the time of their writing in 1848, arguing that capitalism, in comparison to all the "slothful indolence" that went before, is a progressive force in the world. No greater paean to capitalism has yet been written by a capitalist:

> It has accomplished wonders far surpassing Egyptian pyramids, Roman aqueducts, and Gothic cathedrals; it has conducted expeditions that put in the shade all former Exoduses of nations and crusades.
>
> The bourgeoisie cannot exist without constantly revolutionizing the instruments of production, and thereby the relations of production, and with them the whole relations of society. Conservation of the old modes of production in unaltered form, was, on the contrary, the first condition of existence for all earlier industrial classes. Constant revolutionizing of production, uninterrupted disturbance of all social conditions, everlasting uncertainty, and agitation distinguish the bourgeois epoch from all earlier ones. All fixed, fast frozen relations, with their train of ancient and venerable prejudices and opinions, are swept away, all new-formed ones become antiquated before they can ossify. All that is solid melts into air, all which is holy is profaned, and man is at last compelled to face with sober senses his real condition of life and his relations with his kind.
>
> …The bourgeoisie, during its rule of scarce one hundred years, has created more massive and more colossal productive forces than have all preceding generations together. Subjection of Nature's forces to man, machinery, application of

chemistry to industry and agriculture, steam-navigation, railways, electric telegraphs, clearing of whole continents for cultivation, canalisation of rivers, whole populations conjured out of the ground – what earlier century had even a presentiment that such productive forces slumbered in the lap of social labour?

But however admiring, Marx and Engels do not blindly celebrate capitalism. The rest of their oeuvre is devoted to exposing its horrors. Philosopher Steven Shaviro outlines the distinction between the attitude of anti-socialist economist Joseph Schumpeter, who popularised the concept of 'creative destruction' so embraced by contemporary captains of finance, and Marx's position on capitalist dynamism. Schumpeter and the right say that because our grandchildren will be better off, we should just "suck it up." Given how dependent the construction of capitalist society has been upon the primitive accumulation of enclosure and clearances of the commons, upon slavery in the United States, upon the genocide of first nations peoples across the America and Australia, upon colonialism in Africa and India, upon the millions ripped apart on assembly lines or suffocated down mineshafts, and upon the hundreds of millions massacred by the bullets, bombs and gas of imperialist war, the gentle phrase 'creative destruction' is rather so formidable a euphemism as to make Orwell's Ministry of Love appear amateurish. Meanwhile, Shaviro notes, Marx is compassionate to the victims of history, the victims of what French philosophers Gilles Deleuze and Félix Guattari describe as 'deterritorialization'—the constant radical uprooting of fixity under capitalism—*while being "relentlesslessly non-nostalgic."* For Marx, and most socialists after him, there is no going back to the modes of production that went before the current one, which had their own peculiar brutalities of injustice and diabolical scarcity. Instead, the aim must be to "construct a process of advance, but with a managed deterritorialization."[199]

In other words, the point is to take command of this dynamic yet barbaric machine. To retain the vitality of modernity, indeed to unleash its heretofore stifled energies, while doing away with its villainy.

Shouting back to that passage in the *Grundrisse*, Berman reminds us that "Marx wants a truly infinite pursuit of wealth for everyone: not wealth in money—the limited bourgeois form—but wealth of desires, of experiences, capacities, sensitivities, of transformations and developments."[200] Socialism, for Berman, will bring forth so much more of modernity's upheaval: more skyscrapers, more technology, more radical triumphs in art and science and thought, but, crucially, now consciously controlled and democratically organised in the interest of all humanity, instead of unleashed blindly and in the service of the rich minority as occurs under the rule of the market. (It is this latter aspect that distinguishes the socialist modernist from the panglossian liberal, libertarian or militarist cheerleaders of technology, from the young Italian Futurists in 1914 hurling themselves off to "war, the world's only hygiene" through post-war American social scientists' hagiographies of the rhythms of the modern factory and business-management-jargon-spouting 'futurologists' of the 1970s such as Alvin Toffler to today's Randian anarcho-capitalist geek supremacists of Silicon Valley.)

If we can permit ourselves to dream a little, we can go beyond the need to hush the ambient small-is-beautiful rhetoric and defeat the politics of limits for the sake of the scale of ambition necessary to solve the climate and wider 'biocrisis'. We can also say that a rehabilitation of prometheanism is required because to do otherwise requires the human race to abandon its grandest of dreams, of, for example, eliminating disease and of space explo-ration. The first represents a radical reduction in suffering—surely the primary task of any socialist—and the second ensures the survival of our species beyond the life of our sun.

Intriguingly, in Russia in the 1880s, there emerged a school of thought, Cosmism, whose key thinkers such as Konstantin Tsiolkovsky, one of the founding fathers of rocketry and later a Soviet rocket scientist, and Nikolai Fyodorov, who developed the first serious scientific proposals for space travel, believed that it was our species' destiny to settle other worlds and colonise the universe. Though Cosmism was an eclectic salmagundi of science, philosophy and Russian Orthodox mysticism, it ultimately inspired the Soviets' visions of space exploration. Prefiguring today's 'transhumanists', Cosmists also believed that one day, science and medicine would deliver radical life extension and effective immortality. In a 2013 essay on the cosmists, Benedict Singleton, an architect and writer on the politics of technology, captured the essence of their philosophy well: "Storm the heavens and conquer death."[201]

Today, we can extend such lofty but *specific* goals to a more generalised *principle of audacity:* that *our species must continue to achieve ever more impressive feats,* that we must never stop reaching, never stop *progressing*. We, uniquely in nature, have an infinite capacity for ingenuity, what Julian Simon, the libertarian economist and arch-enemy of over-population Cassandra Paul Ehrlich, called the *ultimate resource.* Theoretical physicist David Deutsch explains how we share this phenomenal, open-ended attribute with no other species on Earth:

Using knowledge to cause automated physical transformations is, in itself, not unique to humans. It is the basic method by which all organisms keep themselves alive: every cell is a chemical factory. The difference between humans and other species is in what kind of knowledge they can use (explanatory instead of rule-of-thumb) and in how they create it (conjecture and criticism of ideas, rather than variation and selection of genes). It is precisely those two differences that explain why every other organism can function only in a certain range of environments that are hospitable to it, while humans transform inhospitable environments like the biosphere into support systems for themselves. And while every other organism is a factory for converting resources of a fixed type into more such organisms, human bodies (including their brains) are factories for transforming anything into anything. They are 'universal constructors'.[202]

Cosmologist Stephen Hawking is often quoted as reminding us how *unremarkable* we are: "The human race is just a chemical scum on a moderate size planet, orbiting round a very average star in the outer suburb of one among a billion galaxies." Such vigorous Copernican displacement of the centrality of man in the universe has its uses. It is important that we abandon the essentially religious mode of thinking that assumes that evolution is goal-oriented, that the universe was created for the benefit of us, and even that the universe exists for a reason. Evolution is purposeless and unguided; the universe just *is*. Nevertheless, Hawking is quite wrong with regard to our unremarkableness, argues his colleague Deutsch: we are instead, "chemical scum that dream of distant quasars." We are very remarkable indeed and unlike anything else we've yet discovered in the universe. Our species is just 200,000 years old; our behavioural modernity 50,000. Our age as a scientific civilisation a mere 400–500 years, and the Industrial Revolution is even younger than the Scientific Revolution. In this very brief time, we have achieved great marvels (as well as monstrous horrors, certainly — but, as Stephen Pinker has shown, over the millennia, steadily fewer and fewer of them). Yet we are still only in our infancy when we compare this span of time to the average lifetime of a species, ten million years. Imagine what wonders a ten-million-year-old scientific civilisation will have achieved! To call for a steady-state economy, to oppose growth, is to foreclose all the rest of the spectacular deeds that would otherwise lie in humanity's future.

Moreover, given how robust life is on Earth, how microbes are able to keep on keepin' on happily in some of the most inhospitable environments, from dessicating deserts to the heavy pressures of deep ocean trenches to well below freezing and well above boiling, I have a strong suspicion that simple life is abundant in the universe. I am betting that sometime in the next few decades, we will discover life beyond Earth, either microbial life on Mars or Europa, Enceladus or Titan, or perhaps glimpse the very particular spectroscopic signature of plants' leaves through the atmosphere of distant exoplanets.[203] But as of now, we have a sample size of just one—Earth. For all the hopes of astrobiologists and the billions of others who dream one day of making contact with extra-terrestrial life, we simply don't know yet. If it turns out that in fact life is *not* an intrinsic property of chemical reactivity; if we consistently find no life anywhere else in the universe; if we find dozens, hundreds, thousands of watery, rocky planets in the habitable 'goldilocks' zone around their suns, but it turns out unhappily that they are all barren, then the most profound consequence of this realisation that we are likely the only life in the universe would be that humanity's self-awareness must also be thought of as the universe—nature—becoming *aware of itself*. There is also the possibility that life is abundant in the universe but *intelligent* life is rare or unique. Either way, such a realisation would make the human species all the more precious a thing to behold and take care of. This preciousness would require that we get our act together in terms of assuring an optimal climate for humans and our continued flourishing. But it would also require that we continue to grow economically so that we can, for example, build and maintain effective near-Earth asteroid deflection systems to protect the Earth; spread throughout the galaxy so as to assure the continued existence of the species in the life-vitiating event of a local supernova; and ultimately advance to a level of technology and understanding of reality that perhaps we can figure out a way to permit intelligence to escape the heat death of the universe. We have no reason to believe one way or the other right now that we are the only intelligent life in the universe, but we may as well act as if we are until proven otherwise. And even if the universe turns out to be teeming with intelligent life, we should still want the human race to, ahem, live long and prosper.

Looking forward to what else we can achieve is not to treat scientists, doctors and engineers as the priesthood of a new religion of science or, more prosaically, 'techno-utopianism', as many libel the position. Techno-boosterism blindly optimistically assumes that we need not worry about problems because the right innovation will arrive in the nick of time to save us. The mirror argument, of the neo-Luddites and their precautionary principle, demands that we avoid ever moving forward until we are absolutely sure that it is safe. This blindly pessimistic position is a stance that involves its own set of risks.[204] Between techno-utopianism and neo-luddism there is the aforementioned Promethean optimism that recognises that while at each stage of our history, as a result of our solving past problems, new problems are created, we then must work—and often work very hard indeed—to overcome them. As Deutsch argues, problems are *inevitable*. But problems are also *soluble*.

Over the course of human history, we can see that energy throughput—whether via human or animal labour, wood and dung, fossil fuels or ultimately hydroelectric, solar power and nuclear fission (and one day fusion)—can be seen as a rough proxy for a society's wealth. At each stage, we are solving problems thrown up by the solutions to earlier problems. Along the way, more and more human labour is substituted, increasing productivity; driving up living standards; releasing us from drudgery; curing disease, expanding our understanding of life, the universe and everything; and giving us more time to engage in pleasurable activities. In this way, energy can also be seen as a rough proxy for a society's freedom. As a result of our audacity, our ultimate resource, each of the limits imposed upon us by nature that we have breached—from fire that allowed us to expend less food energy intake on digestion and permitted more energy to be given over to our expanding brain, through electric lighting that allows us to stay up after dark, to the technologies of the bicycle, the washing machine, the pill, abortion, and fertility treatments that have chipped away at patriarchy—has required a growing consumption of energy. All of these natural limits were imposed as arbitrarily as the rules and dictates of any illegitimate government. For this reason, one would think that the most defiant possible demand of anarchism—the political philosophy that challenges not just the power of the state, but all illegitimate authority—would be for the ever greater degrees of freedom delivered by the liberatory power of more energy. Indeed the entirety of the left, not just anarchists, in recognition of this potential for liberation, used to argue not against energy expenditure or technology, but that these advances be shared by everyone, rather than just the elite few.

Energy is freedom. Growth is freedom.

*

We now understand of course that all of this advance has to occur on a sustainable basis—or, more precisely, on a basis that does not inadvertently inhibit open-ended human flourishing—but for the most part, we already know what technologies and organisation of the economy are needed to achieve this sustainability. So while in practice this shift is difficult due to vested interests within capitalism, there is nothing in principle that stops us from achieving it.

Progress and growth need not hold back sustainability. But the politics of limits, like capitalism, certainly hold back progress.

The left needs to recapture the vaulting ambition of idealists like the Cosmists, of the universalist spirit of liberty, equality, democracy and reason of the Radical Enlightenment of Spinoza and Diderot and Paine. We need to embrace once again continent-transforming projects like those of Lenin and Roosevelt. We need to paint on our placards and banners Pankhurst's demand of cornucopian abundance. And, as so many billions around the world did when Neil Armstrong became the first man to step foot on another world, we must learn again how to weep hot tears of pride at the best of what our species can do.

It is time for the left to return, without apology, to the side of progress.

We call ourselves *progressives*, after all, do we not?

Endnotes

1. We called it the Global Justice Movement, while in France it was *Le mouvement altermondialiste.*
2. Naomi Klein, "Capitalism vs the climate", The Nation, 28 November, 2011. http://www.thenation.com/article/164497/capitalism-vs-climate?page=full
3. Naomi Klein, *This Changes Everything,* (New York: Simon and Shuster, 2014), Epub location 146.
4. Klein, Epub location 259.
5. Naomi Klein, "Gulf Oil Spill: A hole in the world," The Guardian, 19 June 2010 http://www.theguardian.com/theguardian/2010/jun/19/naomi-klein-gulf-oil-spill .
6. She develops this idea further in another interview in *In These Times* magazine: "What's clear is that the further we go down this road, and the more this Francis Bacon idea of progress becomes equated with taming and controlling nature, the more these ever-larger and higher-risk technologies are going to take hold." Micah Uetricht, "Naomi Klein: 'We can't dodge this fight' between capitalism and climate change" In These Times, 18 September 2014. http://inthesetimes.com/article/17181/naomi_klein_we_cant_dodge_this_fight_between_capitalism_and_climate_change
7. Klein, Epub location 122.
8. The Dark Mountain Project, http://dark-mountain.net/stories/books/ (7 March, 2015).
9. Ahmed is also the co-director with Dean Puckett of the 2011 documentary *The Crisis of Civilization*: http://crisisofcivilization.com/
10. In particular: Naomi Klein, "Capitalism vs the climate", The Nation, 28 November, 2011. http://www.thenation.com/

article/164497/capitalism-vs-climate?page=full Naomi Klein, "Science is telling us to revolt," New Statesman, 29 October 2013. http://www.newstatesman.com/2013/10/science-says-revolt and Naomi Klein, "Gulf Oil Spill: A hole in the world," The Guardian, 19 June 2010 http://www.the guardian.com/theguardian/2010/jun/19/naomi-klein-gulf-oil-spill (7 March, 2015).

11. Responding To Climate Change, *A systematic approach to climate change* http://www.rtcc.org/2010/html/regione-romagna.html (7 March, 2015).

12. Leigh Phillips, "EU emissions trading an 'open door' for organized crime, Europol says," EUobserver, 10 December 2009, http://euobserver.com/environment/29132 (7 March 2015)

13. David Rich Lewis, "American Indian environmental relations," in Douglas Cazaux Sackman, ed., *A Companion to American Environmental History*, (Hoboken: Wiley-Blackwell, 2010).

14. Quoted in Donald W. Beachler, *The Genocide Debate: Politicians, Academics, and Victims*, (London: Palgrave Macmillan: 2011), 41.

15. Alexander Cockburn, "The Sierra Club's ugly racial tilt", The Albion Monitor, 8 August, 1997. http://www.monitor.net/monitor/9710a/ac-sierraclub.html

16. Catton finished writing *Overshoot: The Ecological Basis of Revolutionary Change* in 1973, but it took him until 1980 to find a publisher.

17. Systemic Capital, Deep Green Resistance: An interview with Derrick Jensen and Rachel Ivey, http://www.systemic-capital.com/deep-green-resistance-an-interview-with-derrick-jensen-and-rachel-ivey/ (7 March 2015)

18. Andrew Simms, "The Impossible Hamster" https://www.youtube.com/watch?v=Sqwd_u6HkMo

19. For those pedants amongst my readers, I will concede that

there are 'caste' divisions within eusocial superorganisms such as bee colonies—organisms made up of other individual organisms. But as the term 'superorganism' suggests, this division of labour and differentiation in consumption rates between, say, queen bees, workers, soldiers and drones, is more akin to the functional speciali-sation and differentiation in metabolic profile between the different organs or cells in your body than to human inequality. Furthermore, individuals of one caste generally lose the ability to perform the characteristic behaviour of the members of another caste, again making this arrangement correspond more to the different roles of the parts of an individual. Finally, the functional specialization of different castes is a behavioural strategy that is driven by instinct, not consciously chosen by an individual or the other individuals within the superorganism, while humans consciously choose the structure (the mode of production) of our society. The discussion as to whether the terms eusociality or superorganism can be used to describe human social organisation as well as that of ants, bees, wasps, termites, naked mole-rats, coral and siphonophores is a fascinating one, and the ontological challenge that superorganisms pose to our conception of individuality is a topic that would go strikingly well with a bong and a bag of Doritos. But none of this contests my essential point here, which is that the scale of lopsided resource distribution and consumption of human class society does not exist anywhere else in nature.

20. Very briefly, as per Marx: the *relations of production* are the sum total of the socio-economic relations in a society. In contrast, the *forces of production* are the combination of our means of labour (technology, infrastructure, land, natural resources, etc.) and our labour power. The relations and forces of production combine together at any epoch as the

mode of production (i.e. capitalism, feudalism, etc.).

21. Friedrich Engels, "Outlines of a critique of political economy", *Deutsch-Französische Jahrbücher* [Paris], February, 1844. https://www.marxists.org/archive/marx/works/1844/df-jahrbucher/outlines.htm

22. Pedants: Let's leave Planck lengths, the shortest meaningful measurable length, aside for the sake of this discussion.

23. Siegfried Franck, et al., "Harvesting the sun: New estimations of the maximum population of planet Earth", Ecological Modelling 12, (June 2011): 2019-2026 DOI: 10.1016/j.ecolmodel.2011.03.030

24. Adjusted by Franck to take into account changes in our understanding of the carbon content of the Earth.

25. I loved *The Road*, gulping the whole deeply miserable thing in one sitting. But it is yet another example of the current fashion for unalloyed hatred and pessimism about humanity. As Rebecca Solnit showed in example after example of human behaviour after catastrophes in her best work, *A Paradise Built in Hell*, the truth is that we are an instinctively cooperative, compassionate species. She rigorously, handily shows that the Hobbesian hypothesis of *Bellum omnium contra omnes* in such circumstances is false. It's a shame that the rest of her oeuvre largely amounts to red-baiting apologism for the abominable Democratic Party.

26. Erle Ellis, "The Planet of no return", *The Breakthrough Journal*, (Winter 2012). http://thebreakthrough.org/index.php/journal/past-issues/issue-2/the-planet-of-no-return

27. Naomi Klein writes for example: "the Breakthrough Institute [is] a think tank that specializes in attacking grassroots environmentalism for its supposed lack of 'modernity,'" *This Changes Everything*, Epub location 98.

28. Denise Brehm, "Scientists decipher 3-billion-year-old genomic fossils", *MIT News*, 21 December, 2010. http://almlab.mit.edu/news.12212010.html

29. Anthony D. Barnosky, et al., "Has the world's sixth mass extinction already arrived?", Nature 472 (3 March, 2011), 51-57. doi:10.1038/nature09678 http://www.nature.com/nature/journal/v471/n7336/full/nature09678.html

30. Gaia Vince, "A looming mass extinction caused by humans", *BBC Future*, (1 November, 2012), http://www.bbc.com/future/story/20121101-a-looming-mass-extinction (7 March 2015).

31. Greg Breining, "Courting controversy with a new view on exotic species", *Yale Environment 360*, (19 November, 2009), http://e360.yale.edu/feature/courting_controversy_with_a_new_view_on_exotic_species/2212/ . (7 March 2015)

32. See for example critical arguments against such comments by Earth First! activist Judi Bari: Beth Bosk, "In the middle of run away history: Judi Bari, Earth First! organiser – Mississippi Summer in the California Redwoods", New Settler 49 (May 1990), http://ecology.iww.org/texts/JudiBari/BethBoskInterview1 (7 March 2015); and by Murray Bookchin in his series of debates with Dave Foreman collected in *Defending the Earth: A dialogue between Murray Bookchin and Dave Foreman*, (Montreal: Black Rose Books, 1999) and in Alan G. McQuillan, ed., *Globally and Locally: Seeking a Middle Path to Sustainable Development*, (Lanham, Maryland: University of America Press, 1998), 31.

33. Naomi Klein, "The change within: The obstacles we face are not just external", *The Nation*, 12 May, 2014. http://www.thenation.com/article/179460/change-within-obstacles-we-face-are-not-just-external

34. Dietmar Post's *Reverend Billy and the Church of Stop Shopping* (2002); Rob Van Alkemade's *Preacher with an Unknown God* (2005) and *What Would Jesus Buy* (2007).

35. Spanish and the indigenous Aymara and Quechua languages are the three official languages of Bolivia.

36. Bob Thomson, "Pachakuti: Indigenous perspectives, *buen vivir, sumaq kawsay* and degrowth," *Development* 54 (2011) 448–454. doi:10.1057/dev.2011.85

37. *The Story of Stuff* http://storyofstuff.org

38. According to research from UK housing and homelessness charity Shelter, half of all adults under 35 will be stuck living in their childhood bedroom within a generation. Admittedly this figure is highly speculative, extrapolating from current trends, but it gives an indication of the scale of the crisis.

39. Anita Singh, "Vivienne Westwood: People who can't afford organic food should eat less", *Daily Telegraph*, 12 November, 2014. http://www.telegraph.co.uk/health/healthnews/11225 326/Vivienne-Westwood-People-who-cant-afford-organic-food-should-eat-less.html

40. Jason Deans, "Jamie Oliver bemoans chips, cheese and giant TVs of modern poverty", *The Guardian*, 27 August, 2013. http://www.theguardian.com/lifeandstyle/2013/aug/27/jami e-oliver-chips-cheese-modern-day-poverty

41. John Powers, "Naomi Klein on *This Changes Everything*, her new book about climate change", *Vogue*, 26 August, 2014. http://www.vogue.com/1009011/naomi-klein-this-changes-everything-climate-change/

42. Victoria Johnson, Andrew Simms, Peter Chowla, *Growth isn't possible*, (London: New Economics Foundation, 2010). http://www.neweconomics.org/publications/entry/growth-isnt-possible

43. William Skidelsky, "Ha-Joon Chang: The net isn't as important as we think", *The Guardian*, 29 August, 2010. http://www.theguardian.com/technology/2010/aug/29/my-bright-idea-ha-joon-chang

44. David Pilling, "Lunch with the FT: Ha-Joon Chang", *Financial Times*, 29 November, 2013. http://www.ft.com /intl/cms/s/2/27a2027e-5698-11e3-8cca-00144feabdc0.html

45. Yasmin Nair, "Why is America turning to shit?", *The Awl*, 30 October, 2013. http://www.theawl.com/2013/10/shit-shit-shit (7 March 2015)

46. United Nations, *Open Defecation*, http://opendefecation.org/ (7 March, 2015)

47. Ethan Siegel, "The physicists' dream machine", *Science Blogs - Starts With a Bang*, 14 August 2013, http://science blogs.com/startswithabang/2013/08/14/the-physicists-dream-machine/#.UgwvF8pSe3c.twitter (7 March 2015)

48. "Treading Lightly", *The Economist*, 19 September, 2002, http://www.economist.com/node/1337251 (7 March 20150

49. Niall Paterson, "Tories going hippy with Happy Planet Index?" 27 February, 2008 *Sky News*, http://news.sky.com /story/573981/tories-going-hippy-with-happy-planet-index (7 March, 2015)

50. Daniel Hoornweg, Perinaz Bhada-Tata, Chris Kennedy "Waste production must peak this century", *Nature* 502, (31 October, 2013), 615-617. http://www.nature.com/news/ environment-waste-production-must-peak-this-century-1.14032#peak doi:10.1038/502615a

51. "NHS is fifth biggest employer in the world", *Daily Telegraph*, 20 March, 2012. http://www.telegraph.co.uk /news/uknews/9155130/NHS-is-fifth-biggest-employer-in-world.html (7 March, 2015). Only three out of the top ten largest firms by number of employees are private companies. The rest are state operations. The top of the list? The US Department of Defense, with 3.2 million workers.

52. Liam Barrington-Bush, "We are not who we were told we were", *Roar*, 1 October, 2014, http://roarmag.org/2014/ 10/naomi-klein-climate-change/ (7 March, 2015).

53. Naomi Klein, "Capitalism vs the climate", *The Nation*, 28 November, 2011. http://www.thenation.com/article/164497/ capitalism-vs-climate?page=full (7 March, 2015)

54. Friedrich Engels, *Socialism: Utopian and Scientific*, (1880) in

Marx/Engels Selected Works, Volume 3, (Moscow: Progress Publishers, 1970), 95-151.

55. Naomi Klein, "Capitalism vs the climate", *The Nation*, 28 November, 2011. http://www.thenation.com/article/164497/capitalism-vs-climate?page=full (7 March, 2015)

56. Naomi Klein, "Climate change, capitalism, and the transformation of cultural values", *Rabble*, 11 November, 2011, http://rabble.ca/columnists/2011/11/climate-change-capitalism-and-transformation-cultural-values (7 March, 2015)

57. Adam Fenderson, "Energy descent action plans – a primer", *Resilience*, 7 June, 2006 (originally published on the now defunct *Eat the Suburbs!* website) http://www.resilience.org/stories/2006-06-07/energy-descent-action-plans-primer-0 (7 March, 2015)

58. The disaster supplies industry alone—freeze-dried food, flashlights, radios, etc.—is $500 million a year in the US, Andrew Martin, "Hurricane Sandy and the disaster preparedness economy", *New York Times*, 10 November, 2012, http://www.nytimes.com/2012/11/11/business/hurricane-sandy-and-the-disaster-preparedness-economy.html?_r=0&pagewanted=all (7 March, 2015)

59. Rebecca Kneale Gould's ethnography of homesteaders, *At Home in Nature: Modern Homesteading and Spiritual Practice in America* (Oakland: University of California Press, 2005), describes the movement thus: "Motivated variously by the desire to reject consumerism, to live closer to the earth, to embrace voluntary simplicity, or to discover a more spiritual path, homesteaders have made the radical decision to go 'back to the land,' rejecting modern culture and amenities to live self-sufficiently and in harmony with nature."

60. National Geographic, *Doomsday Preppers*, http://channel.nationalgeographic.com/channel/doomsday-preppers/ (7 March, 2015).

61. Fellowship for Intentional Community, *Christian Transition Village*, http://www.ic.org/directory/christian-transition-village/ (7 March, 2015).

62. Transition Culture, "An interview with Naomi Klein. Part One: '...that world view is killing us and needs to be replaced with another world view...'" http://transition-culture.org/2011/03/23/an-interview-with-naomi-klein-part-one-that-world-view-is-killing-us-and-that-that-world-view-needs-to-be-replaced-with-another-world-view-%E2%80%9D/ (7 March, 2015)

63. UK Department for Environment, Food and Rural Affairs, "The Validity of Food Miles as an Indicator of Sustainable Development", (London: 2005) http://archive.defra.gov.uk/evidence/economics/foodfarm/reports/documents/foodmil e.pdf .

64. Now, of course if we can genetically engineer luscious, Spanish-style tomatoes to grow in England, we get the best of both worlds. But somehow I reckon the locavores would have none of that.

65. Christopher L. Weber and H. Scott Matthews, "Food-Miles and the Relative Climate Impacts of Food Choices in the United States", Environmental Science and Technology 42 (April, 2008), 3508–3513. http://pubs.acs.org/doi/abs/10.1021/es702969f DOI: 10.1021/es702969f

66. Pierre Desrochers and Hiroko Shimizu, *The Locavore's Dilemma: In Praise of the 10,000-Mile Diet*, (New York: Public Affairs, 2012).

67. C. Saunders, A. Barber & G. Taylor, "Food miles—Comparative energy/emissions performance of New Zealand's agriculture industry." *Lincoln University Research Report No. 285*, (July, 2006).

68. E. Schlich, and U. Fleissner, "Comparison of regional energy turnover with global food." *LCA Case Studies*, (June,

2003), 1–6.

69.   See for example, Susan Krashinsky, "Marketing board failing farmers, study argues", *The Globe and Mail* [Toronto], 28 August, 2012. http://www.theglobeandmail.com/report-on-business/industry-news/marketing/marketing-boards-failing-farmers-study-argues/article4506608/ , A 2008 report to Quebec's minister of agriculture, fisheries and food cited complaints from the popular St-Hubert chain of rotisserie chicken restaurants, that it took "several years ... before we were able to satisfy our clients and offer a 100 per cent grain-fed, air-chilled chicken." "It's ridiculous that it would take so much time and be so arduous," the St-Hubert representative said. "We're currently in the process of getting approval for antibiotic-free chicken. I'll spare you the details of all the hurdles that lie in wait for us in the current system"; and, Sarah Elton, "The 'egg police' crack down on local grey-market eggs', *The Globe and Mail*, 23 February, 2010. http://www.theglobeandmail.com/life/the-egg-police-crack-down-on-local-grey-market-eggs/article1357431/

70.   Jon Fisher, "Global agriculture trends: Are we actually using less land?", The Nature Conservancy, 18 June, 2014, http://blog.nature.org/science/2014/06/18/global-agriculture-land-sustainability-deforestation-foodsecurity/ , (7 March, 2015).

71.   Food and Agriculture Organization of the United Nations Statistics Division, Agricultural indicators – Land, http://faostat3.fao.org/faostat-gateway/go/to/download/ E/EL/E , (7 March, 2015).

72.   Michael Samuels, "Asparagus Rewards Patience," New Hampshire Public Radio 15 May 2014, http://nhpr.org /post/asparagus-rewards-patience .

73.   Naomi Klein, "The change within: The obstacles we face are not just external", *The Nation*, 12 May, 2014. http://www.the

nation.com/article/179460/change-within-obstacles-we-face-are-not-just-external .

74. The shirts are made by Brooklyn arts collective *Not an Alternative* and you can order them here: http://notanalternative.com/blog/goldman-sachs-doesnt-care-if-you-raise-chickens

75. Sasha Lilley, ed., *Catastrophism*, (Oakland: PM Press, 2012), xii.

76. In Jameson's essay, "Future City", he actually says that "someone" said this, but no earlier reference can be found. *New Left Review* 21, May-June, 2003: http://newleftreview.org/II/21/fredric-jameson-future-city .

77. Paul Kingsnorth, "Dark ecology", *Orion Magazine*, December, 2012 http://www.orionmagazine.org/index.php/articles/article/7277 , (7 March, 2015).

78. Franco Berardi, *After the Future*, trans. Arianna Bove, Melinda Cooper, Eric Empson, Enrico, Giuseppina Mecchia, and Tiziana Terranova, (Oakland: AK Press, 2011), 13.

79. Gary Genosko, "Bifo: After the Future", https://vimeo.com/25367464 .

80. Berardi, 107.

81. Alberto Toscano, "The prejudice against Prometheus", *Stir*, Summer, 2011 http://stirtoaction.com/the-prejudice-against-prometheus/ , (7 March, 2015).

82. Mark Fisher, "'They killed their mother': *Avatar* as ideological symptom", *k-punk*, 6 January, 2010, http://k-punk.abstractdynamics.org/archives/011437.html (7 March, 2015).

83. Žižek is here not referring to Communism with a capital 'C', but the earlier, pre-Stalinist communist ideal.

84. Part of the following discussion of the pre-history and advent of *Les Trentes Glorieuses* comes from a much wider discussion of the Eurozone crisis, EU austerity and the various proposals that the left has mounted in response to

the European disaster: Leigh Phillips, "Architectures of a New Europe," EUobserver, 24 January, 2014, http://blogs. euobserver.com/phillips/2014/01/24/architectures-of-a-new-europe/ , (7 March, 2015). For this piece, I rely heavily on the historical research in Sheri Berman's superb 2006 book, *The Primacy of Politics: Social Democracy and the Making of Europe's Twentieth Century* (Cambridge University Press, 2006); the late Tony Judt's masterful tome, *Postwar: A History of Europe Since 1945* (London: Penguin Press, 2005); and Ian Birchall's out-of-print but still essential *Bailing Out the System: Reformist Socialism in Western Europe 1944-1985* (London: Bookmarks, 1986).

85. Götz von Aly, "Die Wohlfühl-Diktatur", *Der Spiegel*, 7 March, 2005, http://www.spiegel.de/spiegel/print/d-39613 406.html (7 March, 2015).

86. James Meek, "Where will we all live?", London Review of Books, 9 January, 2014, 7-16, http://www.lrb.co.uk/v36/n 01/james-meek/where-will-we-live .

87. In 2013, *Red Pepper*, a progressive UK magazine, published on its website a brief article, "Silenced GM scientist speaks out against biotech coercion," about Gilles-Eric Seralini, the French molecular biologist sharply criticised by the scientific community for his infamous and headline-grabbing GMO-rat-tumour study, and promoting his British speaking tour: Claire Walker, "Silenced GM scientist speaks out against biotech coercion", *Red Pepper*, 3 September, 2013, http://www.redpepper.org.uk/silenced-gm-scientist-speaks-out-against-biotech-coercion/. As a science journalist, I was very familiar with the controversy. But I'd also written for the magazine for many years and, while I knew that it leant in the anti-GMO direction, I didn't expect to agree with everything that was published. However, endorsing this discredited quack seemed one step beyond. This is an extended version of the essay later printed in the magazine

that resulted from a subsequent email conversation with the editors.

88. Take the Flour Back, "Take the Flour Back", http://www.you tube.com/watch?v=u5_5dF9Fw8k .

89. Leigh Phillips, "Italian anti-GM group wins destruction of 30-year-old olive-tree project", *Nature News Blog*, 12 June, 2012, http://blogs.nature.com/news/2012/06/italian-anti-gm-group-wins-destruction-of-30-year-old-olive-tree-project.html (7 March, 2015).

90. Virginia Prescott, "The broken olive branch", New Hampshire Public Radio, 18 June, 2012, http://www.nhpr.org/post/broken-olive-branch (7 March, 2015).

91. Leigh Phillips, "Armed resistance", *Nature* 488, (29 August, 2012), 576-579, http://www.nature.com/polopoly_fs/1.1128 7!/menu/main/topColumns/topLeftColumn/pdf/488576a.pd f , doi:10.1038/488576a

92. Leigh Phillips, "Sea versus senators", *Nature* 486, (28 June, 2012), 450, http://www.nature.com/polopoly_fs/1.1089 !/menu/main/topColumns/topLeftColumn/pdf/486450a.pd, doi:10.1038/486450a

93. Oxitec press release: "Oxitec report 96% suppression of the dengue mosquito in Brazilian trials", 21 May, 2013, http://www.oxitec.com/press-release-oxitec-report-96-suppression-of-the-dengue-mosquito-in-brazilian-trials/ , (7 March, 2015).

94. Gruere, Guillaume, Mehta-Bhatt, Purvi, and Sengupta, Debdatta, "Bt cotton and farmer suicides in India", International Food Policy Research Institute, 2008, http://www.ifpri.org/publication/bt-cotton-and-farmer-suicides-india , (7 March 2015).

95. Ballotpedia, "California Proposition 37, Mandatory labeling of genetically engineered food", http://ballotpedia.org/wiki /index.php/California_Proposition_37,_Mandatory_Labelin g_of_Genetically_Engineered_Food_(2012) . (7 March,

2015).

96. Stephen Barrett, "FDA Orders Dr. Joseph Mercola to Stop Illegal Claims", Quackwatch, http://www.quackwatch.com /11Ind/mercola.html

97. http://www.cornucopia.org/wp-content/uploads/2012/08/ prop37-poster.jpeg . (7 March, 2015).

98. "Auchan et Carrefour ont aidé à financer l'étude sur les OGM", L'express, 21 September, 2012, http://www.lexpress.fr/actualite/sciences/sante/auchan-et-carrefour-ont-aide-a-financer-l-etude-sur-les-ogm_116 4587.html , (7 March, 2015).

99. Stephanie Strom, "Has 'organic' been oversized?', New York Times, 7 July, 2012, http://www.nytimes.com/2012/07/ 08/business/organic-food-purists-worry-about-big-companies-influence.html?_r=0 , (7 March, 2015).

100. Verena Seufert, Navin Ramankutty, Jonathan A. Foley, "Comparing the yields of organic and conventional agriculture", Nature 485, (10 May, 2012), 229-232, http:// www.nature.com/nature/journal/v485/n7397/full/nature110 69.html .

101. LUGE (Land Use and the Global Environment) Lab, Navin Ramankutty, http://www.geog.mcgill.ca/~nramankutty/ Research/Research.html (7 March, 2015).

102. Ian Sample, "E coli outbreak: German organic farm officially identified", Guardian, 10 June, 2011, http://www.theguard ian.com/world/2011/jun/10/e-coli-bean-sprouts-blamed

103. Christine Bahlai, et al, "Choosing Organic Pesticides over Synthetic Pesticides May Not Effectively Mitigate Environmental Risk in Soybeans", PLOS One, 22 June, 2010, http://www.plosone.org/article/info%3Adoi%2F10.1371%2Fj ournal.pone.0011250 , DOI: 10.1371/journal.pone.0011250

104. C. Brown and M. Sperow, "Examining the Cost of an All-Organic Diet." Journal of Food Distribution Research 36 (2005): 20–26. http://ageconsearch.umn.edu/bitstream/26759/1/3601

0020.pdf

105. Press release: "Global Organic Food & Beverage Market Is Expected to Reach USD 187.85 Billion by 2019: Transparency Market Research", Albany, 22 October, 2013, http://www.prnewswire.com/news-releases/global-organic-food—beverage-market-is-expected-to-reach-usd-18785-billion-by-2019-transparency-market-research-228748991.html (8 March, 2015).

106. Centers for Disease Control and Prevention, *Current Vaccine Shortages & Delays,* http://www.cdc.gov/VAC CINes/vac-gen/shortages/default.htm#why (8 March, 2015).

107. Centers for Disease Control and Prevention, *Antibiotic Resistance Threats in the United States, 2013,* http://www.cdc.gov/drugresistance/threat-report-2013/ (8 March, 2015).

108. Treatment Action Campaign, "TAC Statement on President Mbeki's AIDS Denialist Remarks in City Press", 26 February, 2006, http://www.tac.org.za/community/node/2226 (8 March, 2015).

109. P. Chigwedere et al, "Estimating the lost benefits of antiretroviral drug use in South Africa", *Journal of Acquired Immune Deficiency Syndrome 49,* (December 2008): 410-5.

110. Robyn Eckersley, "Ecological Intervention: Prospects and Limits", *Ethics & International Affairs* 21, (March 2007): 293.

111. Oil in place: the total hydrocarbon content of an oil reservoir.

112. Michael Raupach et al, "Sharing a quota on cumulative carbon emissions", Nature Climate Change 4, (September 2014): 873-879. http://www.nature.com/nclimate/journal/v4/n10/full/nclimate2384.html doi:10.1038/nclimate2384

113. Robert Peston, *Will China shake the world again?* BBC, 17 February, 2014, http://www.bbc.com/news/business-26225205 (10 March, 2015).

114. Jonathan Kaiman, "China to flatten 700 mountains for new metropolis in the desert", *Guardian,* 6 December, 2012,

http://www.theguardian.com/world/2012/dec/06/china-flatten-mountain-lanzhou-new-area (10 March, 2015).

115. Robert Wilson, Can you make a wind turbine without fossil fuels? *The Energy Collective*, 25 February, 2014, http://theenergycollective.com/robertwilson190/344771/can-you-make-wind-turbine-without-fossil-fuels (10 March, 2015).

116. European Commission DG Environment, *Technologies available to reduce European steel, paper and cement industry emissions*, 28 July, 2011 (Brussels), http://ec.europa.eu/environment/ecoap/about-eco-innovation/policies-matters/eu/585_en.htm

117. Natalie Obiko Pearson, "Cheap electricity for poor squeezing out solar in India", *Bloomberg*, 20 November, 2014, http://www.bloomberg.com/news/2014-11-19/cheap-electricity-for-poor-squeezing-out-solar-in-india.html (10 March, 2015).

118. Alex Trembath, *Grid Governance*, Breakthrough Institute, 16 December, 2014, http://thebreakthrough.org/index.php/voices/energetics/grid-governance (10 March, 2015).

119. Michael Shellenberger, et al, *Our high-energy planet*, Breakthrough Institute, 8 April, 2014, http://thebreakthrough.org/index.php/programs/energy-and-climate/our-high-energy-planet (10 March, 2015).

120. Fred Pearce, "World needs grid power, not just solar panels", *New Scientist*, 4 August, 2014, http://www.newscientist.com/article/mg22329804.400-worlds-poor-need-grid-power-not-just-solar-panels.html#.VB3_Vy5dVT5 (10 March, 2015).

121. John Tarleton, "Naomi Klein breaks a taboo", *The Indypendent*, 12 September, 2014, http://indypendent.org/2014/09/12/interview-naomi-klein-breaks-taboo (10 March, 2015).

122. Der Bundesverband der Energie- und Wasserwirtschaft press conference, "Developments in the German electricity and

gas sector in 2012", 10 January, 2013 http://www.bdew.de/
internet.nsf/id/3EDCA15529136B0AC1257AFD0035A2F6/$f
ile/130110%20BDEW%20Entwicklungen%20der%20deutsch
en%20Strom-%20und%20Gaswirtschaft_englisch.pdf (10
March, 2015).

123. Umwelt Bundesamt, *Greenhouse gas emissions rise again
slightly in 2013, by 1.2 per cent*, 3 October, 2014,
http://www.umweltbundesamt.de/en/press/pressinfor-
mation/greenhouse-gas-emissions-rise-again-slightly-in (10
March, 2015).

124. Indeed, Greenpeace USA's executive director, Annie
Leonard, has been one of the world's leading campaigners
against incinerators.

125. Daniel Weissbach, et al. "Energy intensities, EROIs (energy
returned on invested), and energy payback times of
electricity generating power plants." *Energy* 52 (2013): 210-
221.

126. See also Michael Carbajales-Dale et al., "Can we afford
storage? A dynamic net energy analysis of renewable
electricity generation supported by energy storage", Energy
and Environmental Science 7, (February, 2014) 1538-1544
DOI: 10.1039/c3ee42125b; and Graham Palmer, Energy in
Australia: Peak Oil, Solar Power, and Asia's Economic
Growth, (New York: Springer, 2014). All ably summarised
by Barry Brook of *Brave New Climate* here: http://theenergy-
collective.com/barrybrook/471651/catch-22-energy-storage

127. "Germany's economics minister Gabriel seeks reform for
renewable energy transition", *Deutsche Welle*, 29 December,
2013, http://www.dw.de/germanys-economics-minister-
gabriel-seeks-reform-for-renewable-energy-transition/a-
17329275 (10 March, 2015).

128. "Germany's energy poverty: How electricity became a
luxury good", *Der Spiegel*, 4 September, 2013, http://www.
spiegel.de/international/germany/high-costs-and-errors-of-

german-transition-to-renewable-energy-a-920288.html

129. EU Fuel Poverty Network, *Energy poverty in Germany: Highlights of a beginning debate*, (York: University of York, 7 September, 2014), http://fuelpoverty.eu/2014/07/09/energy-poverty-in-germany-highlights-of-a-beginning-debate/ (10 March, 2015).

130. Will Boisvert, an energy analyst and committed leftist, has written extensively on the problems of the neoliberal *Energiewende*, including in *Dissent*, the American socialist journal: "Green energy bust in Germany", *Dissent*, Summer 2013, http://www.dissentmagazine.org/article/green-ener gy-bust-in-germany , (10 March, 2015); and "The left vs the climate", Breakthrough Institute, 18 September, 2014, http://thebreakthrough.org/index.php/programs/energy-and-climate/the-left-vs.-the-climate , (10 March, 2015).

131. "Sunny, windy, costly and dirty", *The Economist*, 16 January, 2014, http://www.economist.com/news/europe/21594336-germanys-new-super-minister-energy-and-economy-has-his-work-cut-out-sunny-windy-costly , (10 March, 2015).

132. An additional four plants were built in the 1990s for a total of 58.

133. Barry Brook, and Corey Bradshaw, "Key role for nuclear energy in global biodiversity conservation," *Conservation Biology* 0, (September 2014), 1-11. doi: 10.1111/cobi.12433

134. I could spend an entire book debunking the myths of the anti-nuclear movement, but I would just be repeating the excellent work of other progressives such as the solid trade union folks with the Canadian Nuclear Workers' Council, climate scientist and NASA-researcher-turned-environ-mental-activist Jim Hansen, UK left-wing journalist George Monbiot, environmental campaigner and writer Mark Lynas, Californian trade union activist and socialist David Walters of the website *Left Atomics*, Geoff Russell and Barry Brook of the University of Adelaide's Environment Institute

at the Australia-focussed website *Brave New Climate*, and American lefty documentarian Robert Stone in his film *Pandora's Promise*. Liquid fluoride thorium reactor (LFTR) evangelists Kirk Sorensen and Richard Martin are also worth reading.

135. I don't favour the neoliberal boondoggle that is carbon trading or the regressive, flat tax that is carbon taxation that would hope to realise such a carbon price, but this is a useful metric for comparison purposes.

136. International Energy Agency, *Projected Costs of Generating Electricity*, (Issy-les-Moulineaux; 2010).

137. James M. Jasper, "Gods, Titans and mortals: Patterns of state involvement in nuclear development,", *Energy Policy* 20, (Summer 1992): 653-659, doi:10.1016/0301-4215 (92)90007-O; Arnulf Grubler, "The costs of the French nuclear scale-up: A case of negative learning by doing,", *Energy Policy* 38, (September, 2010): 5174–88. doi: 10.1016/j.enpol.2010.05.003

138. Dominique Finon and Carine Staropoli, "Institutional and technological co-evolution in the French electronuclear industry," *Industry and Innovation* 8, (July, 2010) 179–199, doi: 10.1080/13662710120072967

139. Richard Seymour, *Against Austerity*, (London: Pluto Press, 2014), 70.

140. This chapter previously appeared in a slightly different form as an essay in *Jacobin* magazine.

141. Leo Hickman, "James Lovelock on the value of sceptics and why Copenhagen was doomed", *Guardian*, 29 March, 2010, http://www.theguardian.com/environment/blog/2010/mar/29/james-lovelock (10 March, 2015).

142. Ian Sample, "Antibiotic-resistant diseases pose 'apocalyptic' threat, top expert says", *Guardian*, 23 January, 2013, http://www.theguardian.com/society/2013/jan/23/antibiotic-resistant-diseases-apocalyptic-threat (10 March, 2015).

143. Mark Woolhouse and Jeremy Farrar, "An intergovernmental panel on antimicrobial resistance", *Nature* 509, (29 May 2014), 555–557 http://www.nature.com/news/policy-an-intergovernmental-panel-on-antimicrobial-resistance-1.15275 doi:10.1038/509555a

144. Fergus Walsh, "Antibiotics resistance 'as big a risk as terrorism' – medical chief", *BBC*, 11 March, 2013, http://www.bbc.com/news/health-21737844

145. Leigh Phillips, "Socialize Big Pharma", *Jacobin*, 29 June, 2013 https://www.jacobinmag.com/2013/06/socialize-big-pharma/

146. New Drugs for Bad Bugs http://www.nd4bb.eu/

147. Will Steffen, Johan Rockström, and Robert Costanza, "How Defining Planetary Boundaries Can Transform Our Approach to Growth", *Solutions* 2 (May, 2011) http://www.thesolutionsjournal.com/node/935?page=35,0,0,0,0,0

148. Sheila Jasanoff, *Science and Public Reason*, (London: Routledge, 2014) http://www.routledge.com/books/details/9780415524865/

149. Silke Beck, et al, "Towards a Reflexive Turn in the Governance of Global Environmental Expertise. The Cases of the IPCC and the IPBES," *GAIA* 23 (February, 2014): 80–87.

150. Colin Crouch, *Post-Democracy*, (London: Polity, 2004).

151. Valentina Pop, "Eurogroup chief: 'I'm for secret, dark debates'", *EUobserver*, 21 April, 2011, http://euobserver.com/economic/32222 (10 March, 2015).

152. Andy Stirling, *Emancipating Transformations: From controlling 'the transition' to culturing plural radical progress*, (Brighton: STEPS Centre, 2014), http://steps-centre.org/wp-content/uploads/Transformations.pdf (10 March, 2015).

153. Established in 2000 by the then New Labour government of Tony Blair, the commission was shut down in 2011 by the

Conservative-Liberal Democrat coalition under Prime Minister David Cameron.

154. Erik Assadourian, "The Path to Degrowth in Overdeveloped Countries", in *State of the World 2012: Moving toward sustainable prosperity*, (Worldwatch Institute, 2012), http://blogs.worldwatch.org/sustainableprosperit y/wp-content/uploads/2012/04/SOW12_chap_2.pdf (10 March, 2015).

155. Giacomo D'Alisa, Federico Demaria, Giorgos Kallis, eds., *Degrowth: A Vocabulary for a New Era*, (London: Routledge, 2014).

156. As always, abandoning any class distinctions within developed countries. Once again, both my level of consumption and that of Bill Gates are assumed to be identical. Sigh.

157. Serge Latouche, "Degrowth economics", *Le monde diplomatique*, November, 2004, http://mondediplo.com/2004/11/14l atouche (10 March, 2015).

158. Centre for the Advancement of the Steady Sstate Economy http://steadystate.org/

159. John Bellamy Foster, "Degrow or die", Red Pepper, November, 2010, http://www.redpepper.org.uk/degrow-or-die/ (10 March, 2015).

160. Tim Jackson, *Prosperity Without Growth*, (London: Routledge, 2011), 9.

161. US Energy Information Administration, Annual Energy Outlook 2008, (Washington: Department of Energy, 2008) http://www.eia.gov/oiaf/aeo/pdf/0383(2008).pdf

162. For the record, while one can imagine what that policy mix might be, I don't actually think that such policies would be achievable outside the establishment of a genuinely democratic global government.

163. Not absolutely of course, but the problem disappears for a very long time.

164. Or, more correctly, ensuring that we do not damage those external conditions that permit maximum human flourishing.

165. Jonathan Porritt, *Seeing Green: The Politics of Ecology Explained*, (Oxford: Wiley-Blackwell,1984), 44.

166. Joel Kovel, *The Enemy of Nature* (London: Zed Books, 2007), 227–8.

167. Kovel, 229.

168. All this condemning of or ventriloquising for Marx is a bit silly though. One should avoid with all haste anyone vociferously contesting what Marx did or didn't say. Marx's contributions to political economy and sociology are considerable. But he wasn't the Buddha.

169. JB Foster's efforts to insist that Marx was the original green warrior in his 2000 opus, *Marx's Ecology*, extends even to the point of trying to claim that when Marx referred to the "idiocy of rural life" in *The Communist Manifesto*, what Marx had *actually* meant was 'idiot' in the ancient Athenian sense of '*Idiotes*': "a citizen who was cut off from public life." Hmmm. I will readily admit that I have no special hermeneutical expertise to offer here, but upon re-reading that passage, I certainly *like* to think that when Marx—the man who, remember, didn't shy away from describing in print the Reverend Thomas Malthus as *a baboon* and was known for his ten-day benders with his BFF Freddy—talked about rural idiots, he meant what 19th Century readers would have taken the term to mean: rural idiots.

170. Moreover, Foster (and Marx before him) seems to miscomprehend the word 'metabolism' as a synonym for 'exchange', in place of its biological definition, which is simply the collection of all chemical processes in your body that keeps you alive.

171. Paul Hampton spotted this incipient third-worldism hidden in the Rift thesis in his review of *The Ecological Rift*: "Review

of The Ecological Rift by John Bellamy Foster et al",
Workers' Liberty, 21 February, 2011, http://www.worker-
sliberty.org/blogs/paulhampton/2011/02/21/review-
ecological-rift-john-bellamy-foster-et-al (10 March, 2015).

172. Foster actually extends the parable of the soil to *all* human
uses of natural resources, but the critique of 'peakism' holds
regardless.

173. Catherine Clabby, "Does peak phosphorus loom?",
American Scientist 98, (July-August, 2010), 291, http://
www.americanscientist.org/issues/pub/does-peak-
phosphorus-loom (10 March, 2015).

174. Both according to model calculations and direct observation, the IPCC has concluded that global average ozone
depletion has stabilized, with full recovery of the ozone
layer expected in the coming decades, assuming full
compliance with the Montreal protocol: http://www.ipcc
.ch/pdf/special-reports/sroc/sroc_spm.pdf

175. A similar trajectory has occurred with potassium, with
investors opening up new mines in reaction to high potash
prices.

176. Dana Cordell, *The Story of Phosphorus: Sustainability implications of global phosphorus scarcity for food security* (Linköping:
Linköping University, 2010), http://www.diva-portal.org/
smash/get/diva2:291760/FULLTEXT01.pdf

177. No, that's not a typo. McKibben added an extra 'a' to the
name of our planet.

178. "The Hundred Most Influential Books Since the Second
World War", *Times Literary Supplement*, 6 October, 1995.

179. Illich was also a founder of the 'deschooling' or
homeschooling movement (*Deschooling Society* [1971]),
which has in recent years served to give ideological cover to
neoliberal attacks on public education, and launched an
exhaustive attack on modern medicine (*Medical Nemesis*
[1975], AKA *Limits of Medicine*) even as he was afflicted with

a cancerous growth on his face, which he treated with meditation, yoga and opium smoking.

180. David Gibson, "With Philip Blond, age-old Distributism gains new traction", 21 October, 2011, *Anglican Communion News Service*, http://www.respublica.org.uk/item/Age-old-distributism-gains-new-traction

181. Richard Moore-Colyer, "Towards 'Mother Earth': Jorian Jenks, Organicism, the Right and the British Union of Fascists," *Journal of Contemporary History* 39, (July 2004): 353.

182. Graham Macklin, *Very Deeply Dyed in Black: Sir Oswald Mosley and the Resurrection of British Fascism after 1945* (London: I.B. Tauris, 2007), 64–6.

183. Alexander Cockburn, "The Sierra Club's ugly racial tilt", The Albion Monitor, 8 August, 1997. http://www.mon itor.net/monitor/9710a/ac-sierraclub.html   185. http://www.applythebrakes.org/

184. Southern Poverty Law Center, Extremist files – profiles, "Virginia Abernethy", http://www.splcenter.org/get-infor med/intelligence-files/profiles/virginia-abernethy, (10 March, 2015).

185. David Skrbina, "One Vision of Ecological Justice", *The Rewilding Institute*, 27 May, 2013, http://rewilding.org/ rewildit/one-vision-of-ecological-justice/ (10 March, 2010).

186. Janet Biehl and Peter Staudenmaier, *Ecofascism: Lessons from the German Experience* (Oakland: AK Press, 1985). Biehl was the long-time companion of social ecology pioneer, Murray Bookchin.

187. Blake, interestingly, was a friend of the Reverend Malthus.

188. And indeed, the horrors of the USSR.

189. In his 2002 work, *Running on Emptiness*, Zerzan says: "The need for symbols — and violence— ... have their origins in the thwarting and fragmenting of an earlier wholeness, in the process of domestication from which civilization issued." He writes approvingly of the once widely held

position of anthropologists, that Neanderthals did not have language. However, in the last decade, researchers have begun to reassess their position on the linguistic capabilities of Neanderthals and Denisovans—another human subspecies—and pushing back the antiquity of language from the typically quoted 50–100,000 years to half a million years. Intriguingly, this could mean that in our own, contemporary languages, there remains traces of the languages spoken by other hominids. John Zerzan, *Running on Emptiness*, (Los Angeles: Feral House, 2002), 3.

190. Herbert Marcuse, *One-Dimensional Man: Studies in the Ideology of Advanced Industrial Society* (Boston: Beacon Press, 1964), 9.

191. Incidentally, Hölderlin was an early supporter of the French Revolution, planting a "Tree of Freedom" along with comrades from a republican club while a young seminarian (and roommate of Hegel). Despite all I've said about the romantic Counter-enlightenment character of my opponents in this essay, it is of course important to also concede that it is not merely the opponents of rationalism that fall back upon romanticism, but we Enlightenment rationalists, republican revolutionaries and calculating socialist logicians of distribution optimisation also not infrequently draw on the romance of our ideas, lament our own fallen martyrs, and paint fairy-tale pictures of the glorious and just world to come. Even rationalism has a deeply poetic and romantic story to tell of itself. Truth is indeed beautiful. From the Spanish Civil War posters with the anti-fascist slogan '*No pasaran!*' (They shall not pass!) I have on my wall, to that famous anti-Nazi, republican scene in *Casablanca* that gives me goosebumps every time I watch it—where resistance leader Victor Laszlo tells the band in Rick's Café to interrupt the Germans singing *Die Wacht am Rhein* and play the French revolutionary anthem *La*

Marseillaise instead—the Left certainly has its romantic modes of discourse as well. We are all beings pulled as much by pathos as by logos. We are not Spock, *malheureusement.*

192. 'Rootless Cosmopolitan' (*безродный космополит*) was the Stalinist and cryptically anti-Semitic slur against urban intellectuals, artists and scientists thought to be 'anti-patriotic' in the USSR in the 1940s.

193. Contemporary historian Jonathan Israel, perhaps the world's leading scholar on the Enlightenment, usefully makes a distinction between the Moderate Enlightenment of Kant, Voltaire and Hume, figures who made their peace to greater or lesser degrees with monarchy, slavery, religion and the market and the Radical Enlightenment of d'Holbach, Diderot, Condorcet and most especially Spinoza, who extended the supremacy of reason to the full concert of human affairs. Israel shows how the 'complete package' of modern values—individual freedom, democracy, equality, tolerance, freedom of expression, sexual emancipation and the universal right to knowledge and 'enlightenment'—is for the most part ideologically derived from the philosophers of the Radical Enlightenment tradition, not its moderate counterpart. I am very sympathetic to Israel's reading of the two wings of the Enlightenment, although, as ever, history is rarely so neat and tidy. There were many figures that straddled both sides. Nonetheless, Israel's work is a useful, empirical corrective to the standard postmodern narrative.

194. In a wry illustration of the essential conservatism of this fear of advance, of the shock of the new, cartoonist Dresden Kodak casts such dystopian fears back a few ten thousand years in his comic 'Caveman Science Fiction'. One innovative caveman says to the other: "I am make science! I am put fire in cave!" The other caveman, perhaps an early adherent of the precautionary principle, screams: "You go

too far! No control nature!" The first caveman responds: "No! Am keep warm!" and doesn't listen to the warnings of his sage comrade. As a result, the cave is engulfed in flame, and he finally concedes: "Me go too far!" http://dresdencodak.com/2009/09/22/caveman-science-fiction/

195. Karl Marx, Grundrisse, *Foundations of the Critique of Political Economy*, (London: Penguin, 1993), 478.

196. Shaviro's comments here are taken from a lecture he gave in 2013 on the emerging Marxist-influenced philosophical movement that takes the term *Accelerationism*. See Steven Shaviro, "'Accelerationism: An Introduction', Lecture at Grand Valley State University" https://www.youtube.com/watch?v=gi6I0K2PJrw (10 March, 2015), and Shaviro, "Accelerationist Aesthetics: Necessary Inefficiency in Times of Real Subsumption", *e-flux Journal* 46, (June, 2013), http://www.e-flux.com/journal/accelerationist-aesthetics-necessary-inefficiency-in-times-of-real-subsumption/ (10 March, 2015). A glib reading of the Accelerationist philosophical current suggests that its advocates suggest that in order to achieve the radical social change desired, capitalism "must be expanded and its growth accelerated so that its self-destructive tendencies can be brought to their conclusion," a sort of re-hash of 'the worse it gets, the better it gets', or an unconscious aping of Mao's maxim: "There is great chaos under heaven—the situation is excellent." Shaviro, *Post Cinematic Affect*, (London: Zero Books, 2010), 136. I have little time for such hand-rubbing glee at capitalist deterritorialisation. As Shaviro points out, what precisely is the difference between this and the policy desires of the most ardent libertarian? And indeed, this seems to have been the ugly end-point in the trajectory of the philosopher most associated with the movement, Nick Land. Much more interestingly, Alex Williams and Nick Srncek, the authors of the widely shared *Accelerationist*

Manifesto (Critical Legal Thinking, 14 May, 2013, http://criti-callegalthinking.com/2013/05/14/accelerate-manifesto-for-an-accelerationist-politics/), do *not* argue for an acceleration of capitalism for such purposes, but rather for a return to Marx's affirmation of modernity. This latter variety of Accelerationism thus obviously has a great deal in common with my own arguments, albeit having arrived at its conclusions via a philosophical pathway rather than, in my case, a frustration with actually existing green-leftism.

197. Marshall Berman, *Adventures in Marxism*, (London: Verso, 1999), 150.

198. Benedict Singleton, "Maximum Jailbreak", e-flux Journal 46, (June, 2013), http://www.e-flux.com/journal/maximum-jailbreak/

199. David Deutsch, *The Beginning of Infinity*, (London: Penguin, 2011), 58.

200. Sara Seager, et al, "Vegetation's Red Edge: A Possible Spectroscopic Biosignature of Extraterrestrial Plants", *Astrobiology* 5 (March, 2005), 372-390 http://arxiv.org/abs/astro-ph/0503302

Contemporary culture has eliminated both the concept of the public and the figure of the intellectual. Former public spaces – both physical and cultural – are now either derelict or colonized by advertising. A cretinous anti-intellectualism presides, cheerled by expensively educated hacks in the pay of multinational corporations who reassure their bored readers that there is no need to rouse themselves from their interpassive stupor. The informal censorship internalized and propagated by the cultural workers of late capitalism generates a banal conformity that the propaganda chiefs of Stalinism could only ever have dreamt of imposing. Zer0 Books knows that another kind of discourse – intellectual without being academic, popular without being populist – is not only possible: it is already flourishing, in the regions beyond the striplit malls of so-called mass media and the neurotically bureaucratic halls of the academy. Zer0 is committed to the idea of publishing as a making public of the intellectual. It is convinced that in the unthinking, blandly consensual culture in which we live, critical and engaged theoretical reflection is more important than ever before.

ZERO BOOKS

If this book has helped you to clarify an idea, solve a problem or extend your knowledge, you may like to read more titles from Zero Books. Recent bestsellers are:

Capitalist Realism Is there no alternative?
Mark Fisher
An analysis of the ways in which capitalism has presented itself as the only realistic political-economic system.
Paperback: November 27, 2009 978-1-84694-317-1 $14.95 £7.99.
eBook: July 1, 2012 978-1-78099-734-6 $9.99 £6.99.

The Wandering Who? A study of Jewish identity politics
Gilad Atzmon
An explosive unique crucial book tackling the issues of Jewish Identity Politics and ideology and their global influence.
Paperback: September 30, 2011 978-1-84694-875-6 $14.95 £8.99.
eBook: September 30, 2011 978-1-84694-876-3 $9.99 £6.99.

Clampdown Pop-cultural wars on class and gender
Rhian E. Jones
Class and gender in Britpop and after, and why 'chav' is a feminist issue.
Paperback: March 29, 2013 978-1-78099-708-7 $14.95 £9.99.
eBook: March 29, 2013 978-1-78099-707-0 $7.99 £4.99.

The Quadruple Object
Graham Harman
Uses a pack of playing cards to present Harman's metaphysical system of fourfold objects, including human access, Heidegger's indirect causation, panpsychism and ontography.
Paperback: July 29, 2011 978-1-84694-700-1 $16.95 £9.99.

Weird Realism Lovecraft and Philosophy
Graham Harman
As Hölderlin was to Martin Heidegger and Mallarmé to Jacques
Derrida, so is H.P. Lovecraft to the Speculative Realist philoso-
phers.
Paperback: September 28, 2012 978-1-78099-252-5 $24.95 £14.99.
eBook: September 28, 2012 978-1-78099-907-4 $9.99 £6.99.

Sweetening the Pill or How We Got Hooked on Hormonal Birth
Control
Holly Grigg-Spall
Is it really true? Has contraception liberated or oppressed
women?
Paperback: September 27, 2013 978-1-78099-607-3 $22.95 £12.99.
eBook: September 27, 2013 978-1-78099-608-0 $9.99 £6.99.

Why Are We The Good Guys? Reclaiming Your Mind From The
Delusions Of Propaganda
David Cromwell
A provocative challenge to the standard ideology that Western
power is a benevolent force in the world.
Paperback: September 28, 2012 978-1-78099-365-2 $26.95 £15.99.
eBook: September 28, 2012 978-1-78099-366-9 $9.99 £6.99.

The Truth about Art Reclaiming quality
Patrick Doorly
The book traces the multiple meanings of art to their various
sources, and equips the reader to choose between them.
Paperback: August 30, 2013 978-1-78099-841-1 $32.95 £19.99.

Bells and Whistles More Speculative Realism
Graham Harman
In this diverse collection of sixteen essays, lectures, and inter-
views Graham Harman lucidly explains the principles of

Speculative Realism, including his own object-oriented philosophy.
Paperback: November 29, 2013 978-1-78279-038-9 $26.95 £15.99.
eBook: November 29, 2013 978-1-78279-037-2 $9.99 £6.99.

Towards Speculative Realism: Essays and Lectures Essays and Lectures
Graham Harman
These writings chart Harman's rise from Chicago sportswriter to co founder of one of Europe's most promising philosophical movements: Speculative Realism.
Paperback: November 26, 2010 978-1-84694-394-2 $16.95 £9.99.
eBook: January 1, 1970 978-1-84694-603-5 $9.99 £6.99.

Meat Market Female flesh under capitalism
Laurie Penny
A feminist dissection of women's bodies as the fleshy fulcrum of capitalist cannibalism, whereby women are both consumers and consumed.
Paperback: April 29, 2011 978-1-84694-521-2 $12.95 £6.99.
eBook: May 21, 2012 978-1-84694-782-7 $9.99 £6.99.

Translating Anarchy The Anarchism of Occupy Wall Street
Mark Bray
An insider's account of the anarchists who ignited Occupy Wall Street.
Paperback: September 27, 2013 978-1-78279-126-3 $26.95 £15.99.
eBook: September 27, 2013 978-1-78279-125-6 $6.99 £4.99.

One Dimensional Woman
Nina Power
Exposes the dark heart of contemporary cultural life by examining pornography, consumer capitalism and the ideology of women's work.

Paperback: November 27, 2009 978-1-84694-241-9 $14.95 £7.99.
eBook: July 1, 2012 978-1-78099-737-7 $9.99 £6.99.

Dead Man Working
Carl Cederstrom, Peter Fleming
An analysis of the dead man working and the way in which
capital is now colonizing life itself.
Paperback: May 25, 2012 978-1-78099-156-6 $14.95 £9.99.
eBook: June 27, 2012 978-1-78099-157-3 $9.99 £6.99.

Unpatriotic History of the Second World War
James Heartfield
The Second World War was not the Good War of legend. James
Heartfield explains that both Allies and Axis powers fought for
the same goals - territory, markets and natural resources.
Paperback: September 28, 2012 978-1-78099-378-2 $42.95 £23.99.
eBook: September 28, 2012 978-1-78099-379-9 $9.99 £6.99.

Find more titles at www.zero-books.net